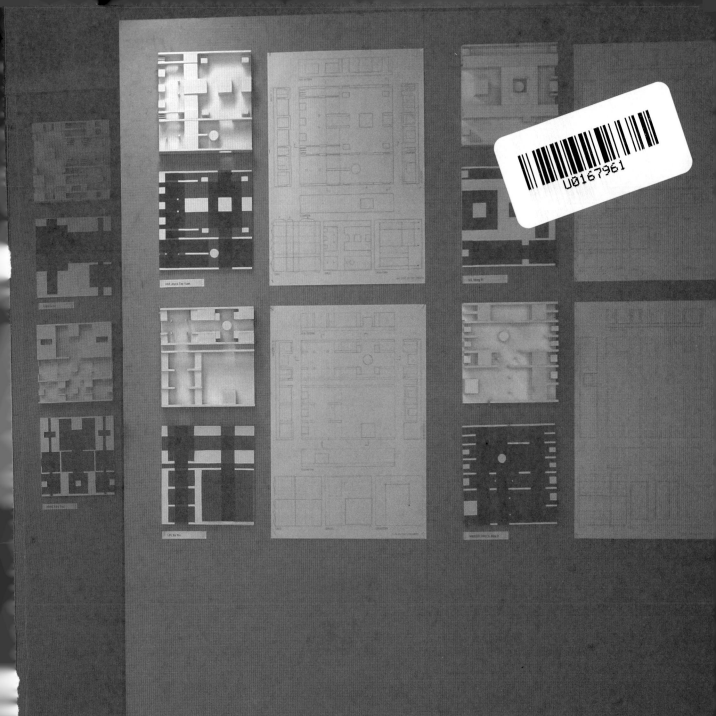

抽象构成与空间形式
Abstract Composition and Spatial Form

[美] 布鲁斯·朗曼　徐亮　著
Bruce Lonnman　Xu Liang

中国建筑工业出版社
China Architecture & Building Press

序言 preface

顾大庆
Gu Daqing

我的同事布鲁斯·朗曼很久以前在俄亥俄州立大学开设过一门设计基础课程，即用装配部件教学法来教授空间构成的形式语言。由于对此的共同兴趣，两人常在一起聊这个话题。有几年我还请他在暑期教师研习班上给老师介绍他的课程。后来我申请了一个中国香港政府的研究经费研究设计教学法，就鼓动他把这个课程的资料整理出来作为一个研究案例，有关的介绍文章发表在《建筑师》杂志上。最近几年我们建筑学院根据大学要求在一年级的通识教育课程中加入一些专业训练的内容，请朗曼来教这个课程。如此，这门课程又被重新激活。徐亮作为助教做了大量课程准备、执行和整理工作。练习的内容有所调整，练习有新的实例。在此基础上我协助策划完成了这本书。

为什么我觉得有必要把该课程介绍给国内的读者呢？主要有以下三个理由：

首先，从设计教学的渊源来说，朗曼是属于"得州骑警"一派，师从李·祝辰、沃纳·塞利格曼和科林·罗。我们以前对"得州骑警"的了解主要是科林·罗、伯纳德·赫伊斯利，以及约翰·海杜克几位"始作俑者"。但是对于其后继的发展，了解的就很少。这本书可以弥补这方面知识的不足。

其次，从教案设计的角度来看，朗曼的课程也是一个很好的范本。该课程最初是大学的一门通识课，泛专业的，严格来说不是一门在设计教室上的设计课。香港中文大学的情况也很相似。课程的性质如此，选课的学生就很多，而老师只有朗曼一人以及没有什么教学经验的研究生助教若干。这对于设计教师来说是一个很大的挑战，因为他不可能采用惯常的设计课做法来应付这个任务。这样的条件下，教学的成功与否完全取决于教案的设计。

最后，这个课程的具体内容——空间练习对国内的读者也很有参考的意义。国内的建筑学校从1980年代中开始在建筑设计基础课程中加入空间练习，即三大构成（平面构成、立体构成和色彩构成）。现在几乎所有的基础课程都有空间练习的内容。但是，我认为大部分这类练习都太直觉，不够严谨，只是一种形式的游戏。而朗曼的空间练习则正好相反，练习的目的不是培养所谓的创造性，而是以一种严谨的方式向学生传授空间构成的基本知识和方法。课程采用了装配部件教学法，即每个练习给予学生一些基本的构件，用来制作符合某种空间秩序原则的空间构成。这些练习从简单到复杂，循序渐进，形成一个结构有序的课程。

总之，我相信无论是对现代建筑教育那段历史感兴趣的学者，还是执教基础课程的老师以及建筑学的初学者都能从研读本书的过程中获益良多。

My colleague, Mr. Bruce Lonnman taught a beginning design course at The Ohio State University long time ago. The course is about the abstract language of space-formation based on the method of "kit-of-parts". Due to our mutual interest, we often exchange our views on this topic. I also invited him several times to give a lecture about his course at the NSBAE Workshop on Design Pedagogy. Later, I got a research grant from the Research Grants Council of the Hong Kong Special Administrative Region on design pedagogy. Taking his course as a case study, I encouraged him to put his material into an article, which was eventually published in the journal *The Architect*. A few years ago, our school introduced a new design related course in the first year general education package as requested by the university. Mr. Lonnman was asked to teach this course. In this way, this old course has been reactivated. Mr. Xu Liang has assisted Mr. Lonnman in course preparation, execution and documentation through these years. The course contents have been readjusted to the new teaching context and new exercise samples have been collected. Based on these works, I have helped to arrange the publication of this book.

There are three reasons for me to introduce this course to the mainland readers.

First, in terms of the origin, Mr. Lonnman is obviously a follower of the Texas Rangers, learned directly from Lee Hodgden, Werner Seligmann and Colin Rowe. We have been well informed about the achievements of those pioneers such as Colin Rowe, Bernhard Hoesli and John Hejduk, but know very little about the works from their students. This book could be a reference on the later development of this group.

Second, Mr. Lonnman's course can be served as an exemplar of program design for design teachers. The course was originally designed as a general education course opened to students from different academic backgrounds in The Ohio State University. Strictly speaking, it was not a studio course taught in studios. The teaching context at CUHK is quite the same. Due to this nature, Mr. Lonnman usually has to teach a large class of students with only the help of a group of unexperienced graduate students. This is a great challenge for the course teacher as he cannot cope with the task using the conventional studio method. Under such condition, the success of the course solely depends on the design of the program.

Third, our Chinese readers might find their great interest about the content of the course — exercises on space perception and composition. The kind of abstract exercises on space composition was introduced to the mainland architectural schools in the mid of 1980s known as Sanda Goucheng (three courses on abstract form-making). It becomes quite popular in almost every architectural foundation course today. However, from my point of view, most of these exercises are too intuitive, lack of rigor and merely formal games. Mr. Lonnman's course is just the opposite, its main objective is not to cultivate creativity but to teach students the knowledge and method of space composition in a rigorous manner. Adopting the "kit-of-parts" method, in each exercise, students are asked to make space composition with a set of given elements, which demonstrates a principle of spatial order. These exercises form a logic development from simple to complex. Along with model-making exercises, students also learn the basics of graphic presentation.

All in all, I do believe that either scholars who are interested about the history of modern architectural education, or design teachers who would like to learn the method of kit of parts approach, or architectural beginning students who would like to master the power of space organization, all can benefit from reading and studying this book.

前言 foreword

布鲁斯·朗曼
Bruce Lonnman

建筑这一学科与绘画、雕塑，甚至是音乐并无多大的差别。它们都要求从抽象的角度思考，并使用一套专业的语言系统进行交流。设计的过程包含了灵感、技巧以及一定程度上的历史记忆。这些艺术形式均重视经典作品的价值，它们作为范例指明了什么是可实现且经久不衰的，它们构成了指导当代创作的重要数据库。不考虑过往知识的积累则如同白费力气去重复工作。建筑是不断革新又同时建立在过往基础上的一门学科。抽象、语言、先例与技巧组合在一起，共同贡献于建筑设计这项技艺的学习。

我们相信，按照一个结构有序、系统的过程来教授建筑学是非常必要的，它可以为学习设计（尤其是在初始阶段）提供最为清晰且目标明确的方法。在设计过程中，建筑师往往需要对包括场地、客户要求或气候在内的一系列情况作出回应。若换做是白纸一张，所面对的设计挑战便会大大增加：看似拥有一切可能，却似乎又无从下手。没有了诸如场地、功能需求的限制，可以自由创作的设计师对于要做什么又会少些方向。如果用一组细致考量过的限制条件作为指引，设计概念的生成也会变得相对容易。

本质上来说，设计是一个解决问题的过程，它时刻与形式发生关联。本书收录的练习均与此想法有关，每个练习都在探讨一个或多个与建筑设计相关的形式问题。这些练习旨在培养学生对于空间限定的敏感性，并推崇丰富且易读的空间状态。在课堂中，我们严格控制各练习的训练时长，要求学生在限定的"场地"内，满足特定的"使用计划"，通过使用一组给定的"要素"来创造一个合适的形式解决方案。这里所说的要素指的就是"装配部件"，这个"部件"或要素是预先确定好的，我们以此来降低设计方案的形式复杂程度。而同时，为设计创意的发挥提供足够的空间，这一点从丰富多样的最终设计成果里可以得到证实。

在这本书中，我们分享了一些开展的基础设计课程，或者说"建筑设计预备工作室"的经验。课程的重点在于空间构成的基础，我们通过设计这些练习来促进对形式与秩序基本原则的学习，并发展学生进行抽象设计的能力。同时，安排一系列的课程讲座与这套练习配套，从中向学生介绍各种技巧、解释建筑的基本原则，并引用一些建成的及未建成的建筑案例佐以说明。这门课程的另一重要目标是与学生展开关于设计的对话。这些对话是否有意义，很大程度上取决于相应词汇及语法的建立。以此为基础，学生方能清晰准确地表达出其设计想法。

The discipline of architecture is not so different from that of painting, sculpture, or for that matter, music. Each requires the ability to think in abstract terms and to communicate through a language specific to the discipline. The process of design involves a combination of inspiration, technique, and, to some degree, historical memory. Every art form has a tradition of classic works serving as models of what is achievable and enduring. These models or precedents constitute a vital database that informs the present. Discounting the accumulated knowledge of the past would be akin to reinventing the wheel. Architecture is evolutionary while building on the past. Abstraction, language, precedent, together with technique, contribute to learning the craft of architectural design.

We believe that teaching architecture as a structured and systematic process is not only necessary, but that it provides the most objective and transparent method of learning design, particularly at the beginning stage. To design is to react to context whether it is the site, the client's brief or the climate. There is nothing more challenging than a blank canvas. Everything is possible except knowing where to start. Without limitations such as site boundaries or program requirements, the designer can do anything but has little guidance on what to do. Forming a design concept becomes a much less formidable task when guided by a set of carefully established constraints.

All of the exercises in this catalogue accept the idea that design is essentially a problem solving process that is integral with form. Each exercise is intended to address one or more formal issues relevant to architectural design. All exercises share a common goal of developing sensitivity for spatial definition while, at the same time, encouraging an appreciation for both complexity and legibility. Many of the exercises are time-constrained, in-class design exercises

with "site" boundaries, specific "program" requirements, and a restricted set of "elements" with which to create an appropriate formal solution. The elements used in the designs are referred to by a familiar term; "kit-of-parts". The "parts" or elements have been pre-determined so as to minimize the formal complexity of a design solution. Yet there is ample room for invention; the variety and range of solutions to each exercise supports this fact.

In this book we share the experience of teaching basic design as a "pre-architecture design studio" course. The emphasis is on the fundamentals of spatial composition and the exercises have been created to facilitate learning basic principles of form and order, and to develop the ability to work with abstraction. Supporting these design exercises is a series of lectures that introduce technique, explain the fundamental principles of architecture and draw parallels to both built and un-built works of architecture. Perhaps the most important goal of the course is to initiate students into a conversation about design. And to be meaningful, this conversation like others requires vocabulary and grammar, as well as ideas that can be articulated with clarity and precision.

目录 contents

3 练习 exercises

4 回探 backdrop

PLAN LIBRE

1 课程　program

1.1 架构 framework

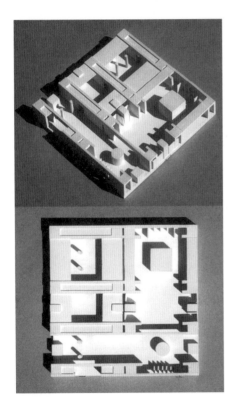

平行墙练习习作;
Example of Parallel wall project;

本书呈现的基础设计课程有5个主要组成部分，按照主题依次分为技法、形式、构成、系统与体验，每一主题对应安排有系列讲座与设计练习。每一部分还可细分为多个知识要点，它们是设计教学中重点探讨的一些问题。

"技法"部分旨在介绍建筑的图示再现方法：工具、绘图技巧、建筑图的类型以及设计过程中绘图的作用。这一部分的讨论从制图工具的介绍开始，然后是识别建筑再现中普遍使用的主要图示类型。训练的重点在于徒手线条的绘制，要求线条清晰可读、控制精准。绘图在设计过程中的不同作用，比如说信息记录或者促进设计想法的变形与发展，会在讲座"绘图与过程"中详细介绍。

"形式"部分开启了对设计与过程的讨论。对于形式的认识来自于实体与虚空、体量与容积的对应关系，这有助于将空间理解为一种可塑、可视、可描述的三维形式。继而，空间的问题就是对形状与尺寸的形式处理，这一点贯穿于课程全部的设计练习之中。

课程中，我们称空间构成的形式想法或基本组织概念为设计组织策略构思。识别设计中的组织策略，就是将不同的设计按照基本的结构或模式进行分类。由此，我们可以对采用相似构思的不同设计方案进行比较，并以单一图解表述不同的设计可能。形式可以通过一系列的加法、减法操作进行生成与变形。建筑先例已为我们概念性地呈现了这些基本操作方式，学生需要在设计练习中进一步探索：通过组织一组给定的要素来创造一个实体体量并体现减法或加分操作的形式品质。

模数概念（即具有相对比例关系的一组尺寸）是"装配部件"方法[1]的典型特征，该方法被广泛应用于本书介绍的大部分练习中。明确且清楚定义的限制条件，对于基础设计训练而言是有益的。从这个角度来说，装配部件既是一种操作限制，也是一种为形式生成提供指引的"支架式"[2]机制。在"实体"与"容积"这两个练习中，装配部件就是一组预先设计的

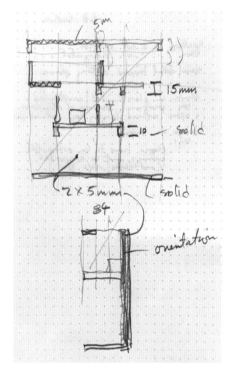

剖面空间生成练习设计笔记；
preparation notes for Spatial development in section；

具有模数关系的要素，这些要素需要全部使用在设计中。

第三部分"构成"审视了一组可赋予形式以特点、限定秩序的形式组织原则。首先，通过理解点、线、面之间的关系，可以建立起对"空间清楚限定"的认识。如何使用最少的点来限定一个立方体容积？如何用不同的板片边界组合来强化空间的限定，并渗透诸如正面性与层级关系的设计特征？通过参照建筑先例，对于空间限定特点的认识可以得到进一步强化。

建筑形式与秩序的五个基本组织原则（轴线、对称、层级、重复与基准）可通过引用一些建筑与城市的案例[3]进行详细说明。对于这些组织原则的深入理解是十分必要的，这是发展出一套设计语言，并以此对形式进行批判性讨论、创造出丰富而不损失可读性与视觉秩序的构成设计的基础。

这一部分的第三个要点是"图底关系"。"图底关系"源起于格式塔心理学的相关研究，心理学家鲁道夫·阿恩海姆[4]以此为基础展开了一系列的考察。在建筑中，"图底关系"现象作为一个概念，被认为是设计图示与分析的有力工具。它作为三维现实的一种二维抽象，有助于理解"实体为图"或"空间为图"的含义。在建筑与城市设计中，还可以使用这一概念来剖析与构成相关的诸多问题，包括图形空间设计，图与底的关系，剖碎的层级化解读以及现象的透明性。

在"系统"部分，空间秩序与结构之间的相互关系被认为是一种概念的秩序建构手段。网格所暗示的是结构单元的复制，在此着重探讨均跨与非均跨之间的空间构成，平行与正交的空间区域，网格场地与物件的组织关系，墙体与框架，格点处与网格间的空间规划等。

"基准墙"练习从一个墙体物件的形式特点出发，考虑网格构成的生成。练习中对于基准墙的回应主要体现在场地上各部件的组织，这包括限定高度的线性墙体与一组部件要素。通过布置这些"主角"以寻求网格构图的解决方案，并以此验证基准墙之于整个方形场地的特定模数关系。

The program in beginning design presented in *Abstract Composition and Spatial Form* is comprised of five main sections that correlate with the lectures and exercises. The topics are: technique, form, composition, system and experience. Each section contains several related sub-topics that focus on important issues explored in the studio sessions.

Technique provides an introduction in architectural graphic representation: tools, drawing technique, types of drawings and the role of drawing in design process. It begins with a description of the tools for drafting and then identifies the main drawing types commonly used in architectural representation. An emphasis is placed on freehand drawing technique with the goal to develop legibility, precision and proportion in sketching. The different roles of drawing in the design process, such as recording information or facilitating transformation and development of an idea, are explained in the lecture, Drawing as Process.

Form initiates the discussion on design and process. The comprehension of form as both solid and void, mass and volume is seen as the first step in the cognition of space as a three-dimensional entity that can be configured, visualized and described. This concept of space as a form possessing shape and dimension is intrinsic to all of the exercises in this program.

The formal idea or basic organizational concept of a spatial composition is referred to in our lessons as a parti. Identifying the parti of a design is a way of categorizing the scheme according to its basic structure or pattern. This enables comparison with other schemes of similar parti revealing the variations possible within a single diagram. Form can be generated and transformed through the processes of addition and subtraction. These basic operations are presented conceptually through architectural examples and then explored in design exercises using prescribed modular elements to create an object that has qualities of subtracted or additive form.

The modular concept of a set of proportionately related dimensions is an important feature of the kit-of-parts[1] approach that is adopted in most of the exercises in this book. Beginning design exercises can benefit from clear, defined parameters and limitations. In this sense the kit-of-parts is both a limitation as well as a liberating *scaffolding*[2] device that offers suggestive guidance in generating form. In the exercises Mass and Volume, the kit-of-parts is a pre-defined set of modular elements that must be used in each design.

The third section, *Composition*, examines the formal principles that characterize form and define order. First, the prerequisite of clear space definition without ambiguity is established through an understanding of point-line-plane. What is the minimum number of points that defines a cuboid volume? How can planes and edges combine in various ways to enhance space definition and imbue properties such as frontality and hierarchy? These and other characteristics of space definition are reinforced with reference to architectural precedents.

Five basic principles of architectural form and order (axis, symmetry, hierarchy, repetition and datum) are defined and contextualized through architectural and urban examples[3]. A deep understanding of these principles is viewed as essential in developing a vocabulary with which to critically discuss form and to create compositions that achieve complexity without sacrificing legibility and visual order.

The third sub-topic in this section is figure/ground. Derived from studies in Gestalt psychology and examined in the writings of Rudolf Arnheim[4], the figure/ground phenomenon is a concept that has

期末评图之前，2014-2015学年学生；
students before final review, AY 2014-2015;

简单地说，网格表达的是正交方形单元的重复，它是无层级关系的、均质的。通过变形可以将它转换为一个有差别的、有层级关系的网格模式，这就蕴含了处理不规则、非对称的功能关系与场地情况的可能。在此，我们测试了两种不同的网格结构："网格框架"练习是由约翰·海杜克的"九宫格问题"发展而来，开间数增加到五跨；"平行墙"练习中使用的则是一个更加密集的平行承重墙系统，由此带来了对方向性与层级性的讨论。

课程的最后一部分"体验"，提醒我们空间设计的基本目的：作为人类可栖居的场所，需要积极的回应自然力量，并通过尺度的控制与使用者的身体产生关系。在"剖面空间生成"中，通过限制楼层高度、引入室内采光的研究，将先前的抽象形式训练过渡到关乎人体尺度、功能性的设计问题的考虑，并引入对于"氛围"概念的理解。通过关注人在设计之中的位置与视线来展开讨论关于尺度、空间感知与自然采光等的一系列问题。这些问题开始在抽象的设计练习与关乎人居的现实的空间设计之间投射。

1　装配部件是一种启发式教学机制，可见于得州骑警所创造的设计练习中，该教学思想的讨论详见"方法：装配部件教学法"一节。
2　"支架式"一词是指教育研究中发展出的一种学习技巧，它鼓励通过适当的提问或指示在解决问题的过程中提供指引。
3　秩序建构的诸原则最初由罗杰·舍伍德在其名为《建筑的原则与要素》的著作中提出。
4　感知心理学家、教育家鲁道夫·阿恩海姆著有《艺术与视知觉》（1954）一书，该书探索了我们对于形式感知的心理学基础及影响。阿恩海姆对于图底关系这一视觉现象的描述与解释对于建筑教育有着重要作用。

found powerful application in architecture as a tool for visualization and analysis. As a two-dimensional abstraction of three-dimensional reality, the figure/ground is essential in understanding the meaning of 'object as figure' or 'space as figure'. It provides a valuable insight into many compositional aspects of architecture and urban design including figural space design, field and figure relationships, hierarchical interpretation of poché, and phenomenal transparency.

In *System*, the relationship between spatial order and structure is introduced as a conceptual ordering device. Grid presupposes a condition of structural unit bay repetition and explores the resulting impact on spatial composition of equal versus unequal bays, parallel versus cross-grain spatial zoning, grid field and object relationships, wall versus frame, and on-grid versus off-grid spatial planning.

The Datum Wall exercise considers the generation of a grid organization from the formal characteristics of an object wall. Reacting against the datum wall is the creation of a field structure consisting of linear wall elements of fixed height and a set of object elements. Together these protagonists seek resolution in a grid composition that justifies the idiosyncratic modulations of the datum wall with the overall square site.

At its most basic level, grid assumes an orthogonal, square bay repetition. Non-hierarchical and neutral, the transformation of this a priori condition leads to differentiated, hierarchical grid patterns capable of accommodating unequal and asymmetric program and site conditions. Two examples of grid structure are tested. The Grid Frame, a 5x5 bay descendant of the Hejduk Nine Square problem, and the Parallel Wall, a more dense system of parallel, load bearing walls that have directionality and hierarchy.

The final section of the program, *Experience*, reminds us of the prime objective of space design as a habitable place of man respon-

sive to the forces of nature and related by scale to the physical dimension of user. In the project Spatial development in section, floor-to-floor dimensions and the introduction of an interior lighting study moves the exercise away from the earlier abstract formal exercises towards a design problem with human scale, hinting at functionality and the notion of *ambience*. The position and viewpoint of an occupant in the design introduces questions concerning scale, spatial perception, and natural lighting. These new issues begin to inform the abstraction of the design exercises with the reality of the design of space for human habitation and delight.

1   Kit-of-parts is a heuristic device adopted in some of the design exercises by the Texas Rangers. The historical evolution and application in architectural education is explored in the essay, "Pedagogy: The Kit-of-parts Approach".
2   The term "scaffolding" refers to the name of a learning technique developed in educational research that recommends the introduction of questions or statements at appropriate intervals to provide guidance in solving a problem.
3   This formulation of ordering principles was initially developed by Roger Sherwood in the publication entitled, *Principles and Elements of Architecture*.
4   Rudolf Arnheim, perceptual psychologist, educator and author wrote the book, *Art and Visual Perception* in 1954. The book explored the psychological basis and context that informs our perception of form. Arnheim's description and explanation of the figure-ground phenomenon was influential on architectural education.

1.2 关于本书 about the book

《抽象构成与空间形式》一书记录了一门建筑学基础设计课程的教学。课程由一系列讲座、阅读材料及设计练习组成，它旨在介绍建筑形式与构成的基础问题。讲座主要介绍诸如空间限定、实体与虚空、形式秩序原则、网格组织等主要概念，它们是设计教学的概念支撑。每一个练习都是一个独立的设计问题，它贡献于理解空间及其在建筑中的核心作用。

本书由三部分组成：讲座、练习以及回探（由三篇阐释教学方法源起与背景的论文组成的附录）。第一部分为讲座内容的总结，平行于设计练习设置，旨在为练习中所探索的概念提供历史与理论基础。讲座中选取了建筑、景观与城市的相关案例来说明基本问题与原则。同时，对于建成作品的讨论架构起联系抽象设计练习与基本设计原则的实际应用的桥梁。本书的第二部分囊括了至今为止经教学实践过的各设计练习。除了设计任务书外，我们还提供了一些学生习作的范例以及一则反思各设计练习教学目标的小结。

基础设计课程的教师可能会认为本书所记录的设计练习太过局限或粗浅。事实上，每一个问题都是一个限定明确的设计练习，一个在相对短时间内完成的关于形式的设计研究。然而，随着教学的深入，新的练习会在前面练习所讨论知识的基础上结合新的设计问题。以此，训练的复杂性递增，学生也需要更多的时间来完成设计。在接近尾声的时候，诸如平行墙或网格框架的设计练习则需要更多课时来发展设计，消化教师的评价与建议并制作最终模型、绘制图纸。

我们希望本书所介绍的这种基础设计教学方法能够适用于不同的教学大纲或教学目标。它的特点在于平衡结构有序的教学与个人创意力的探索。我们相信每个设计练习都为学生提供了足够的创意表达机会，同时在一个共同的基础上做到有目的地比较与评价学生作业。我们相信经过该课程，学生可以掌握一种建筑设计工作方法——通过一连串想象、测试、讨论、修正的过程最终完成设计成果的表现，并了解基本的建筑原则。

Abstract Composition and Spatial Form describes a course of instruction for beginning architectural design students. It outlines a program of study comprised of lectures, key reference readings and design exercises that introduce the fundamentals of architectural form and composition. The lectures present thematic topics (space definition, solid and void, formal ordering principles, grid organization, etc.) that support the studio work outlined in the exercises. Each exercise is an independent design problem focused on a particular issue that is key to an understanding of space and its essential role in architecture.

There are three sections in this book: the *lectures*, the *exercises* and the *backdrop* — an appendix with three essays to explain the origins and elaborate on the background of the teaching methodology. The first section contains the notes to the lectures that run parallel to the studio and provide a historical and theoretical basis for the ideas explored in the design exercises. Examples of buildings, landscapes and cities are selected to illustrate fundamental issues and principles. The discussion of actual built work serves as a bridge between the abstraction of the design exercises and the application of design principles in the real world. In the second section, the book offers the statements or briefs for all of the exercises that have been tested in our teaching to date. In addition to the assignment handout, we offer examples of student work and a segment on reflection that is intended to identify the goals and objectives of each exercise.

Instructors of beginning design courses should not be concerned that the exercises described in this book appear too limited or rudimentary. Each problem is a *constrained* design exercise; a didactic formal design study intended to be completed in a relatively short time. However, as the exercises increase in complexity, combining knowledge from previous exercises while introducing new issues, they require more time. Toward the end, problems like *Parallel Wall* or *Grid Frame* require several studio periods for design development, instructor criticism and guidance, as well as additional time for the making of the final model and drawings.

We hope that you will find this method of teaching beginning design applicable to your curriculum or educational objectives. Our basic philosophy in this approach is to achieve a balance between structured learning and the pursuit of individual creativity. We believe that each design exercise provides sufficient opportunity for creative expression while at the same time, establishing a common basis on which each student work can be objectively critiqued and compared against many alternative schemes. Through the process of imagining, testing, discussing, revising and finally, the presentation of their work, we believe beginning design students acquire a method of working as well as an exposure to the fundamental principles of architecture.

2 讲座　　lectures

2.1

技法［绘图］
technique [drawing]

平面, plan　　　　屋顶平面, roof plan

正立面, front elevation　　侧立面, side elevation

AA剖面, AA section　　BB剖面, BB section

建筑再现：技法与惯例
关键概念：建筑图的类型（正投形，平行投形与灭点透视），
绘图规则，线型，平面、立面、剖面，正面与立面，比例，比
例尺，轴测，水平斜等轴测与正面斜等轴测。

　　以图纸与模型作为建筑再现的主要媒介，常被建筑师用
于推敲设计方案、与客户沟通设计、指导建筑物的准确施工
等。作为建筑设计的最后一个阶段，在施工中常涉及到与大
量负责建造实施的承建商、施工人员之间的沟通。这就要求
有详细的文件来说明建筑的每一个细节，也就是施工图纸。
图纸的绘制需要遵循一套预先设定的规则，并以此控制图纸
的连续性与可读性。
　　最基本的三种建筑图是<u>正投形</u>，<u>平行投形</u>与<u>灭点透视</u>。正
投形图是对物体或建筑进行扁平、二维的再现，其中包括平面、
剖面与立面。平行投形呈现的是三维的视图，包括水平斜等轴
测、正等轴测与正面斜等轴测。透视图是最真实的一种建筑再
现方式，它包括一点透视、两点透视与三点透视。[图2.1-1、2.1-2]
　　将建筑的某单面向与之平行的垂直面进行投形，所获得
的视图即为立面图，投形方向与该面正交垂直。获得视图的
投形平面可以看作是一个置于建筑前方、与建筑该面相平行
的一个剖切面。所有的建筑都可以按照方位罗盘的基本方向
（东南西北）对应确定四个基本侧面（或者说立面）。一栋
建筑通常包含五个立面：四个侧面与一个屋顶平面。也就是
说，"屋顶平面"就是屋顶部分的立面视图。和其他的正投
形图一样，立面图的绘制同样具有等比例关系。在立面图
中，需要按照一个特定的<u>比例尺</u>来绘制垂直与水平边界、建
筑各特征等。比如说，如果将一个实际高10m的山墙屋脊按照
1:100的比例尺绘制其立面图，它在图纸上的高度应该是
10cm，也就是以1cm来表达100cm或1m。
　　对建筑进行水平剖切，移走剖切面上方的部分，向剖切

正面斜等轴测（60°），
elevation oblique (60º)

一点透视，
1-point perspective

图 2.1-2
平行投形图与灭点透视图；
paraline and perspective
drawing;

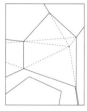

正等轴测（45°–45°），
axonometric (45º-45º)

两点透视，
2-point perspective

正等轴测（30°–30°），
isometric (30º-30º)

三点透视，
3-point perspective

architectural representation: technique and convention

Key concepts: *drawing types (orthographic, paraline and perspective), drawing conventions, line weight, plan, elevation, section, frontality and façade, proportion, scale, axonometric, plan and elevation oblique.*

Architectural representation in the form of drawings or models is the medium that architects use to study and develop a design, to communicate the design to a client and to direct with precision the construction of the building. The last phase of a building design, the construction phase, requires communication with numerous contractors and builders who are responsible for the realization of the built design. This phase requires detailed documents referred to as construction drawings that describe the building in every detail and adhere to a set of prescribed conventions that inform the consistency and legibility of the drawings.

The three basic types of architectural drawings are *orthographic*, *paraline* and *perspective*. Orthographic drawings are flat, 2-dimensional representations of an object or a building. They include plan, section and elevation. Paraline drawings present a 3-dimensional view and include axonometric, isometric, elevation oblique. Perspective drawings are the most realistic representation of architecture. They include 1, 2 and 3-point perspectives. [fig. 2.1-1, 2.1-2]

An elevation is an orthogonal view (perpendicular to the building surface) of the side of a building projected onto a vertical "picture" plane. The projection plane can be envisioned as a cutting plane positioned in front of and parallel to the side of a building. All buildings have four principle sides or elevations corresponding to the cardinal compass points (North, East, South and West). A building generally has five elevations: the four sides and the roof. A "roof plan" is an elevation of the roof. Elevations, like all orthographic projected drawings, are scale drawings. Vertical and horizontal edges and features of the building are drawn in an elevation at a particular *scale*. For example, a gable roof ridge that is 10 meters high drawn at a scale of 1:100 will be 10cm in the elevation (1cm = 100cm or 1m).

A plan is a view of a horizontal section cut through a building (with the top portion removed), looking down. All solids that are cut by this imaginary cutting plane are drawn in profile (the outside edge of the cut solid) with a heavy dark line that is called a section or profile cut line. Important edges of openings or mezza-

图 2.1-3
沙姆伯格住宅，
理查德·迈耶，1972-74，
二层平面，剖面；
Shamberg House,
Richard Meier, 1972-74,
2/F plan, section;

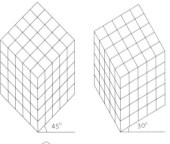

图 2.1-4
正等轴测图：45°/45°，
30°/60°，30°/30°；
45° - 45° axon, 30° - 60° axon and
30° - 30° isometric;

面下方望去所获得的视图就是平面图。在平面图中，被这个假想的剖切面所剖切到的实体部分应该使用粗实线（亦称为剖切线或轮廓切线）来绘制其轮廓（剖切实体部分的外沿边界）。如果这个剖切面上方存在开口或夹层楼板，则需要在平面图中以虚线表示其主要边界。

若对建筑在垂直方向进行剖切，则可以获得剖面。剖切面的选择通常以最大限度地表达设计信息为目标。剖切线或剖面轮廓线是室内空间的轮廓，也是建筑外部可感知形状的轮廓。剖面图的重点在于表现空间及其形状或形式，其中对空间轮廓的呈现最为重要。在剖面图中，顶棚及墙体内部的材料信息通常是不予表现的。场地剖面图通常会剖切到建筑两侧的地面，以展现建筑与场地之间的关系，包括植被、景观特点以及地面的轮廓形态，而建筑的细节会相对简化。

沙姆伯格住宅 [图2.1-3] 的上层平面图展现了线宽控制的准确与细致。假想的剖切面所切到的要素用深色的剖切轮廓线来表示。墙体及柱的内部进行留白处理，对应建筑空间的布局则由深色的剖切轮廓线描绘。另一种绘图方式是将剖切到的墙柱等要素内部涂黑或是通过排线（等距平行排布的线条）填充。沙姆伯格住宅的剖面图中，垂直剖切面所在的位置剖切经过连桥、入口以及主要的通高空间，即设计的主要特征。同时，将平面与剖面对位摆放，有助于建立起关于空间布局的三维认知。

平行投形表现的是一种实测图。在实际的物体或者建筑中，相互平行的线条（边界）在平行投影图中依旧保持平行。所有的线条（边界）应当与图纸的X-Y-Z坐标系平行，并按照一定的比例尺（如1:100）绘制。水平斜等轴测（即轴测图）的绘制是将该物件或建筑的平面沿垂直方向、按照其换算高度投形而获得的一种平行投形图。正面斜等轴测则是将物件或建筑的某一立面沿进深方向进行投形获得。正等轴测则是以一个变形后的平面（其上物件或建筑的直角不再保持90°）作为基准面，与图纸坐标系相平行的各线条（边界）应按照同一比例尺绘制。

在轴测图中，通常会对物件或建筑进行一定的旋转，角度在0°到90°之间。如果采用"45°/45°"，最终会呈现为沿对角线对称、两侧立面在变形与视觉层级上一致的视图；如果采用"30°/60°"，其中一侧面会产生较小的变形，其正面性的呈现更多，视觉的层级性也就更高 [图2.1-4]。在詹姆斯·斯特林设计的英国剑桥大学历史系大楼的图纸 [图2.1-5] 中，使用的是"45°/45°"轴测图。这一角度有助于表现建筑本身的几何特征：主阅览室沿45°对角线对称。就正面性而言，这

图 2.1-5
英国剑桥大学历史系大楼，
詹姆斯·斯特林，1963–67，
模型，轴测图；
History Faculty Building,
Cambridge University,
James Stirling, 1963-67,
model, axonometric drawing;

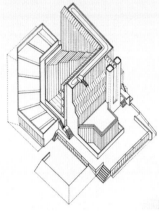

nine floors that are above the cutting plane should be indicated with a dashed line.

A section is a vertical cut through a building. The cutting plane of the section is usually taken at the place where it will show the most information. The cut line or section profile line is an outline of the interior space. It is also an outline of the perceivable shape of the exterior of the building. A design section emphasizes the shape or form of space. The profile of the space is primary. Information about the material inside the ceiling or wall is not drawn. Site sections cut through the ground on both sides of a building show more of the relationship of the building to the site. The building detail may be simplified in a site section. Trees, landscape features and the contour of the land are shown.

The upper level plan of the Shamberg House [fig. 2.1-3] notes the precision and the careful control of *line weight*. Elements cut by the imaginary cutting plane are indicated with a dark section profile line. The interior of walls and columns are left white while the configuration of space is delineated by dark section profile lines. An alternative graphic approach is to "fill in" or "poche" the interior of walls, columns, etc. with solid black or line hatching (closely spaced parallel lines that fill these areas). In the section of the Shamberg House, the cutting plane for this section is located to cut through the bridge, entrance and main double height space, all of the important features of the design. Here the alignment of the plan and section is beneficial in developing a 3-D understanding of the spatial configuration.

A paraline drawing is a measured drawing. Lines (edges) that are parallel in the actual object/building are parallel to each other in the paraline drawing. All lines (edges) that are parallel to the primary axes (x, y, z) are drawn at some measured scale (e.g. 1:100). The plan oblique (axonometric) drawing is a paraline drawing in which the vertical dimension of the object/building is projected up from a true plan. In the elevation oblique, the lateral depth of the object/building is projected back from a true elevation drawing. The isometric drawing is a projection up from a distorted plan (right angles in the actual object/building are no longer 90º). However, the length of all lines (edges) parallel to the primary axes is dimensionally accurate in the scale they are drawn.

The orientation of the object/building in an axonometric drawing may be rotated to any angle between 0º and 90º. If the

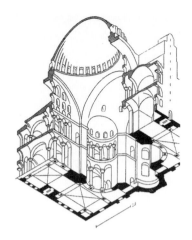

图 2.1-6
圣索菲亚大教堂仰视图，
奥古斯特·舒瓦齐，1899;
Worm's eye view of Hagia
Sophia,
Auguste Choisy, 1899;

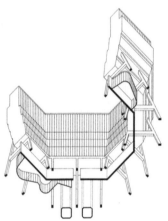

图 2.1-7
弗洛里大楼，
詹姆斯·斯特林，1966-71，
仰视轴测图;
Florey Building, The Queen's
College, Oxford,
James Stirling, 1963-67,
worm's eye axonometric
drawing;

个"45°/45°"轴测图是中性的，两侧面的变形均等，没有哪一侧相对突出。这一视图倾向于将建筑作为一个物体，以提供一个全局的、多角度的呈现。

正等轴测也是一种平行透视图。它与斜等轴测之间的差别在于：正等轴测所使用的平面是经过变形的，变形时需控制平面各边与水平方向之间的夹角。如果这个物件或建筑的底面是正方形（即四角均为90°），其各边在典型的正等轴测中会保持与水平方向30°的夹角。正如所有的平行透视图一样，正等轴测图的绘制同样需要按照相同的比例尺来确定物件各边长。通过平面沿垂直方向投形，并按照同一比例尺换算垂直各边的高度。当然，正等轴测图中各底边与水平方向之间的夹角可以是任意的。若按照"15°/15°"绘制，会产生一个扁平、铺展的视图，视点会显得较低；而如果按照"45°/45°"绘制，就是平面不变形的斜等轴测图。

奥古斯特·舒瓦齐绘制的圣索菲亚大教堂 [图2.1-6] 常作为插图出现在建筑设计相关的历史与理论专著中。在此，建筑被提升、旋转，以提供一个由下向上的视图。这种类型的视图被称作是"仰视图"。整个建筑被穿过中央穹顶的垂直面剖切，并在底部对应绘制平面。整幅图的迷人之处正是其一体地表现建筑外部形式与内部空间体量、平面与剖面，它是"全视图"的最初始形式。建筑师詹姆斯·斯特林也曾使用这样的绘图方法来表现他的设计作品 [图2.1-7]。它已经成为了斯特林的一种标志性图示。

正面斜等轴测是第三种平行透视图。正如其名称所暗示的，绘图时利用无变形的正立面作为基准来生成三维视图。由该立面向其后深度方向沿投形线投形，常见的投形角度为30°。视图中各线条的长度按对应物件的实际长度，按统一的比例尺换算。这样，虽然除了正立面之外，其他各面均发生了一定程度的变形，但是图中各尺寸及其比例关系仍是准确的。

史蒂芬·霍尔是一位在作品图示中广泛使用正面斜等轴测的当代建筑师，其斯特列多住宅图解 [图2.1-8] 就是很好的例子。除了采用正面斜等轴测的制图方法外，该图纸还进一步抽象：将建筑按照形式系统区分为几部分，并在图示中进行分离。在地面层绘制有建筑底座以及作为服务的狭长的砌体墙体块；飘浮其上的是玻璃隔断墙与外部玻璃围合；顶部的是弯曲的屋顶板片。这些要素都是按照准确的尺寸与对应的位置绘制，只是按照一定的距离进行垂直提升。

右侧是古意大利山顶小镇圣吉米尼诺亚的一张照片以及12世纪早期描绘这座小镇的壁画 [图2.1-9]。艺术家想象着鸟瞰

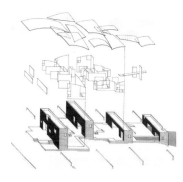

图 2.1-8
斯特列多住宅，
史蒂芬·霍尔，1989-91，
分层轴测图；
Stretto House,
Steven Holl, 1989-91,
axonometric drawing;

图 2.1-9
上：意大利圣吉米尼诺亚；
下：《海边城市》，安布罗焦·
洛伦采蒂，约1340年；
up: view of San Gimignano,
Italy;
down: City by the Sea,
Ambrogio Lorenzetti, c.1340;

angles are 45° and 45° there is a diagonal symmetry and both side elevations are equal in distortion and visual hierarchy. If a 30° - 60° orientation is chosen, one of the sides will have less distortion and will have more visual hierarchy; it will have more frontality. [fig. 2.1-4] The drawing of the History Faculty Building design [fig. 2.1-5] by James Stirling at Cambridge University in England is a 45° - 45° axon. This orientation favors the geometry of this building that has a 45° diagonal symmetry of the main reading hall. The 45° - 45° axon view is neutral with respect to frontality: neither side has prominence and both are equally distorted. This view has the tendency to visually emphasize the building as an object, to be seen in-the-round equally from all sides.

The isometric is also a paraline drawing. The difference between an axonometric and an isometric is that in an isometric drawing, the plan is not true, that is, it is distorted by the chosen angles of the orientation of the base to the horizontal. If the base of the object or building is square (the corner is 90°) then these angles will be equal. A typical isometric projection will have the edges of the plan drawn parallel to 30° angles. As in any paraline drawing, all edges that are parallel in the original object are drawn parallel (and in scale) in the isometric. Vertical edges are projected vertically and their length is drawn to scale. The isometric drawing can use any angle of rotation to the horizontal. Angles of 15° and 15° will produce a flat and spread out view with the viewpoint appearing to be lower. Note that the choice of 45° - 45° will produce an axonometric in which the plan is not distorted.

A drawing of the Hagia Sophia in Constantinople by Auguste Choisy used as an illustration in an historical and theoretical treatise on architectural design [fig. 2.1-6]. The building here is lifted up and rotated to provide a view from below looking up. This type of drawing is referred to as a "worm's eye" view. In addition, the building is sectioned vertically through the central dome and the bottom of the building is drawn as a plan. The beauty of this phenomenal drawing is that one can view simultaneously the exterior surfaces, the plan, the section and the interior spatial volume. It is the original "all in one" view. The architect James Stirling adopted this particular drawing type in representing his own work [fig. 2.1-7]. For Stirling it became a signature drawing.

The elevation oblique is the third type of paraline drawing. As the name implies, the drawing uses an elevation, drawn frontally in

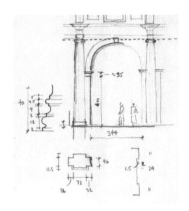

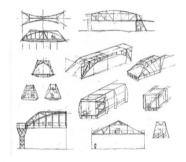

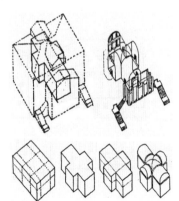

这座被墙包围的小镇，并以一系列独立、不连续的正面斜等轴测图来将其进行三维描绘。它试图从一个不可到达的高处进行俯瞰来重建一个真实而再现的小镇视图，它的出现比灭点透视的制图法早了一个多世纪。

最后一种三维再现图示是透视。它是物件或建筑描绘中最真实的一种再现方式，与我们的视觉观察方式最为接近。灭点透视制图技法是在15世纪初，文艺复兴时期"重新发明"的。如同立面与剖面一样，透视依赖于假想的视图平面，一个用以接收透视投形的表面。绘制一点透视图时，需要建立一个视点、一个消失点及一条水平线。此外，透视图还包括有两点透视与三点透视两种。

过程

关键概念：测绘图，设计思考，分析，表现图，施工文件，设计过程，假设，设计任务书，表现测试，设计概念。

在建筑中，图纸既是一种探索工具，也是一种记录方法。在过去，图纸是记录并沟通建筑知识的主要手段。在中世纪的行会，速写本是记录设计与施工信息的珍贵资源。在现存最早的一本速写本中，意大利建筑师佛朗契斯科·迪·乔吉奥（1439-1502）绘图记录了他拜访过的建筑，内容关于机器及其他设计发明的想法，以及比例的理论性研究等。

建筑师主要在以下四种情况下绘图。如前面所述，其中第一种是记录、存档。比如说某现存建筑物的测绘图纸，或是场地踏勘时的一些绘图记录。它们包括徒手绘制的一些视图、根据测量绘制的平面图解或场地剖面 [图2.1-10]。

第二种与绘图相关的活动是"设计思考"，在此绘图是一种想象与推演的手段。草图或者徒手画研究贯穿于从概念到最终形式的各个设计阶段，它有助于建筑师推敲形式并从视觉的角度思考。在研究设计想法与不同形式可能时，设计师会使用不同的图纸类型，如平面、剖面、透视等 [图2.1-11]。

绘图同样是形式分析时所使用的一种基本工具 [图2.1-12]。分析就是甄别与挑选设计特点的一个过程，它涉及到观察与慎思。就记录而言，相机无疑是一种非常有效的工具，它接近于人眼的视觉观察。虽然十分准确，但它是不加筛选且中性的。分析的目的在于抽取其中有意义的特别信息，并在这个探索的过程中展现出对作品形式组织与视觉表现的深层理解。

图纸的第四种用途是再现或沟通设计概念、草案或最终设计，它的对象可以是客户、施工承建商或者合作的建筑师等。如果对象是客户之类的非专业人士，就需要使用到表现图：偏

the picture plane with no distortion, as the generator of the three dimensional view. An angle such as 30° is selected to project lines from the elevation into the depth of the background. The length of these lines is dimensional and corresponds to the actual length (in the real object) by the scale of the drawing. Thereby the drawing is dimensional and to scale, although distorted on all surfaces except the elevation that is frontal.

Steven Holl is a contemporary architect who has adopted the elevation oblique on many occasions. The drawing of the Stretto House [fig. 2.1-8] is a good example. In addition to being an elevation oblique, the drawing also employs a further abstraction, that of pulling apart sections of the building to visually separate the different formal systems. At the ground level the building plinth and the masonry walls of the narrow service cores are drawn. Floating just above are the glass partition walls and exterior glass enclosing walls. At the very top are the curved roof planes. These elements are all drawn in their correct dimension and orientation but are lifted vertically by a fixed dimension.

A view of the ancient Italian hilltop town known as San Gimignano and an early 12th century fresco of the town [fig. 2.1-9]. The artist imagined this aerial view of the walled town and drew it three-dimensionally as a series of independent and discontinuous elevation oblique views. It was an attempt to give a realistic and representational view of the town from an inaccessible vantage point more than a century before the invention of perspective drawing.

The last category of three-dimensional representation is the perspective. The perspective is the most realistic representation of an object/building. It is the closest approximation to how we see. The technique of perspective drawing was developed in the Renaissance in the early 15th century. Filippo Brunelleschi may have been the first architect to use perspective for a building design. The perspective, like an elevation or section, relies on an imaginary picture plane that represents the surface of the perspective drawing. Establishing a station point, a vanishing point and the horizon line enables the construction of the perspective. There are several types of perspectives, each involving one, two or three vanishing points.

process

Key concepts: *measured drawings, design thinking, analysis, presentation drawing, construction document, design process, hypothesis, program brief, performance testing, design concept.*

Drawing in architecture is both a tool for discovery as much as a method of recording. In the past, drawings were the primary means of capturing and communicating knowledge about architecture. During the era of the medieval guilds, the sketchbook was an invaluable source of information about design and construction. In one of the earliest sketchbooks in existence, the Italian architect Francesco di Giorgio Martini (1439 – 1502) made drawings of buildings he visited, ideas for machines and other inventions, theoretical studies of proportions, and much more.

Architects use drawing in four basic ways. As mentioned, the first is to *record* or *document*. For example, measured drawings of an existing built work. Or drawings made on a site visit. These may include freehand studies of views, a diagrammatic plan or a constructed site section based on measurements [fig. 2.1-10].

A second activity involving drawing is *"design thinking"*. Drawing is a means to imagine and speculate. Sketches or freehand drawing studies enable a designer to generate form and visually think through the stages of transformation from concept to final form. The designer uses many different types of drawings (plan, section, perspective, etc.) in studying the idea and various formal alternatives [fig. 2.1-11].

Drawing is an essential tool in the *analysis* of form [fig.2.1-12]. Identification and selection of specific characteristics of a work is an analytical process that involves observation and critical judgment. The camera is a useful tool for recording something in a way that closely approximates our vision. It is precise, but non-selective and neutral. To analyze is to extract particular information that has meaning. It is a process of discovery that reveals a deeper understanding of the formal organization and visual expression of a work.

A fourth use of drawing is to *represent* or *communicate* a design concept, scheme or final design to either a client, a contractor, a building official or another architect. These drawings are typically referred to as presentation drawings if they are for non-technical persons such as a client. A preference is for less abstract and more realistic rendered views. Communication with the building

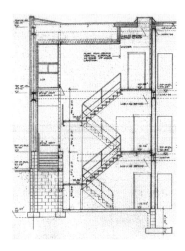

图 2.1-13
施工图纸（作者绘制）；
construction document
(by author);

具象的、写实的渲染图。通常会使用施工图纸与施工承建商进行沟通，采用的是标准的制图方法，尺寸标注及相关信息 [图2.1-13]。在未来，这些信息可以通过三维电脑建筑信息模型来进行交流。

什么是设计过程？过程指的是为准备或制作某物时所进行的一系列有所指引的动作。传统认为，设计与制作是紧密相连的，实际加工工艺影响了设计师及其设计决策。任何设计过程都涉及解决问题。科学方法指的是17世纪以来，为有效探索自然科学中的某些现象所建立的程序，它包括五个步骤或阶段：问题的阐述，研究与分析，提出假设，验证假设，在证据的基础上接受或修正假设。

设计遵循一条类似的轨迹。项目往往从关于设计内容的问题开始：一个房间、一栋建筑还是一座城市。问题的范畴连同各种条件参数、限制以及所遇到的特殊情况，一并构成了问题的阐述。其中的一部分内容可以通过与客户商议确定，另一部分则需要在准备列名各空间需求的设计任务书的过程中予以完善。在一些情况下，场地必须是确定的。接下来，建筑师就要研究设计问题中他所不熟悉的部分，比如说场地性质、建筑类型、特别而具体的功能要求、使用者的需求等。实际上，研究阶段可以是无穷无尽的，设计者会在这个阶段针对设计问题进行形式研究，并尝试不同的解决可能。这些设计研究就等同于"假设"或推断，以决定采纳或否决。通常会就此提出初步的设计方案，并以此为基础做进一步深化。不论是哪个阶段，可以通过评价与表现测试（模拟）来对设计进行"检核"。这个阶段所获的信息会反映在设计的调整中。在一定时候，确认设计的最终方案，通过使用图纸、模型与客户沟通其中细节，以得到批准并提交承建方准备施工。

设计的过程是迭代的、非线性的。虽然直觉有一定的作用，但是不断对可行方案进行更新与严格评价的目的性研究是一种普遍被接受的设计方法。在这个过程中，生成设计概念的技巧与途径多种多样。作为决策项目特点与概念想法的设计师，其独特的定位是至关重要的。

contractor is usually through a set of construction documents that are conventionalized architectural drawings with dimensions and notes [fig. 2.1-13]. In the future, this information will be communicated in a digital three dimensional *building information model* (BIM).

What is the design process? A process implies a guided series of actions in the preparation or making of something. Traditionally, design was closely connected to the making of an object. The actually crafting of the thing being made informed the designer and influenced the design. Any design process must include problem solving. The scientific method is the well-established procedure for investigating phenomena in the natural sciences followed since the 17th century. It involves five steps or stages: Statement of the problem, research and analysis, hypothesis, testing the hypothesis and confirmation or revision of the hypothesis based on the evidence.

Design follows a similar trajectory. A design project begins with a problem to design something: a room, a building, or a city. The scope of the problem together with parameters, boundaries and special conditions to be met, constitutes the problem statement. This is partly done in consultation with the client and in the preparation of a program *brief* that identifies the space requirements. Sometimes a site must be determined. Next the architect will research the unfamiliar aspects of the problem including the nature of the site, the type of building, specific and detailed functional requirements and the needs of the user, and much more. In fact the research phase can be boundless. The designer during this phase will develop formal studies as alternate solutions to the design problem. These alternative design studies are the equivalent of "hypotheses" or conjectures that must be proven or rejected. A design scheme is proposed at this point and developed in more detail. At every stage the design will be "tested" by a combination of critique and performance testing (simulation). The knowledge gained through this phase will inform and modify the design. At some point the design will be accepted as a finished design and the detailed characteristics of the design will be communicated through drawings and models to both the client for approval and the contracted builders for construction.

The process of design is iterative and non-linear. Intuition plays a role but objective enquiry through the generation and rigorous critique of potential design schemes is the most widely accepted method of design. Within the process, the techniques and approaches for generating a design concept vary greatly. They will depend on the particular orientation of the designer, who must determine the character and conceptual idea of the project.

2.2

形式［要素］
form [elements]

图 2.2-1
卡夫拉金字塔，
约公元前2530年；
Pyramid of Khafre, ca. 2530 B.C.;

图 2.2-2
阿特留斯宝库，剖面，平面；
Treasury of Atreus,
section, plan;

实体与虚空

关键概念：理想实体，轮廓，朝向，加法、减法。

实体与虚空，或由此推导而来的体量与容积，它们之间的对比是建筑形式的本质特点。在自然界中，实体与虚空具体表现为山体与洞穴。两者都是有形状的。山体的形状是可见的，是以天空为背景的实体轮廓。通常情况下，它是凸形的。而另外一方面，洞穴的形状是其内部虚空部分的轮廓。虚空本不可见，只能凭靠想象。洞穴内壁的凹面如同一个模具，它呈现出洞穴的容积形式。在不同位置对洞穴内壁进行剖切，可以捕捉到这个容积的轮廓。

人类文明早期的纪念物在本质上都是实体体量，其内部有时会存在细碎的小空间。古埃及金字塔［图2.2-1］是一种<u>理想实体</u>，形式上它是一个规则多面体。金字塔体量共有五面，其中四面为三角形，以及一面方形底面。它是集中性的，形式的对称暗示了位于正中央穿过底面方形的中心及锥顶的垂直轴线。垂直向上是它唯一的朝向。与此相对的另一个墓葬纪念物是阿特留斯宝库（约公元前1350年），它仅有一个可体验的内部容积，其外部是与内部空间形式无关的土堆［图2.2-2］。墓腔的内部平面为圆形，剖面上是尖拱形。在坟墓周围需要布置一个正式的入口大厅以进入到这样的一个空间，使之贯穿拱形坟墓并明确朝向。

古代建筑中，几乎所有实体体量的纪念物都是采用砌体施工方法：将石块切割、凿刻成特定的几何形状，然后彼此堆叠以创造体量形式。这样的施工操作是做<u>加法</u>。阿特留斯宝库就是采用这种加法的施工方法。另一方面，埃及方尖碑［图2.2-3］是一种雕塑体量：将大块的石头削切、凿刻以将其塑造成最终的形式。它可以认为是一个<u>减法</u>的操作过程。雕塑建筑的一个特别案例是位于埃塞俄比亚拉利贝拉的圣十字教堂（约12世纪晚期）［图2.2-4］。这座如同从火山岩中挖出的一个

图 2.2-3
弗拉米尼奥方尖碑，于公元前10年搬至罗马；
The Flaminian Obelisk, relocated to Rome in 10 B.C.;

图 2.2-4
圣十字教堂，拉利贝拉；
Church of St. George in Laibela;

solid and void

Key concepts: *platonic solid, profile, orientation, additive, subtractive.*

The contrast between solid and void or its corollary, mass and volume, is an essential characteristic of architectural form. In nature, solids and void are present as mountains and caves. Both have shape. The mountain's shape is seen as a profile of a solid form against the sky as a background. It is mainly convex. The shape of a cave, on the other hand, is the profile of a void, something not seen and only imagined. The inner concave surface of the wall of the cave is a like a mold revealing the cave's volumetric form. Sections cut at different points through the wall of the cave reveals the profile of the void.

The earliest monuments of man were essentially solid masses, sometimes with minor voids existing within. The great pyramids of Egypt [fig. 2.2-1] are *platonic solids*, forms that have sides or facets that are regular polyhedrons. The pyramid is a five sided mass, four of which are triangles and the fifth a square base. It is centric, meaning that the symmetry of its form implies a central vertical axis from the center of the square base through its apex and beyond. The pyramid has no orientation other than upwards. In contrast, another burial monument, the Treasury of Atreus (ca. 1350 B.C.) has only an interior void that can be experienced; the exterior is a mound of earth whose shape is unrelated to the form of the space below [fig. 2.2-2]. The interior void is circular in plan and an ogival arch in section. To enter such a space requires a formal entrance hall that penetrates the vaulted tomb on some point of its perimeter, thus establishing orientation.

Almost all of the solid mass monuments of ancient architecture were of masonry construction; blocks of stone cut and chiseled to a specified geometry and then stacked upon each other to create an object form. This process of construction is *additive*. The Treasury of Atreus is such an additive construction. An Egyptian Obelisk [fig. 2.2-3] on the other hand is a sculpted mass. A very large stone is cut and chiseled, reshaping it into its final form. This might be thought of as a process of *subtraction*. A unique example of a sculpted building is the Church of St. George in Laibela Ethiopia (ca. late 12th century) [fig. 2.2-4]. Literally carved out of volcanic rock the church building sits as a cruciform object in a large and deep trapezoidal void.

Addition and subtraction can also refer to the perceived char-

23

图 2.2-5

圆厅别墅, 安德烈亚·帕拉第奥, 约1570年,
轴测图, 体量关系图解;
Villa Rotondo,
Andrea Palladio, ca. 1570,
axonometric drawing, massing
diagram;

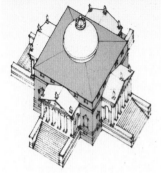

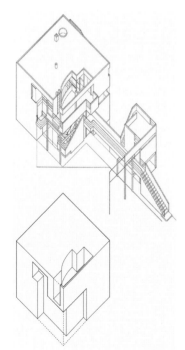

图 2.2-6

汉索曼住宅,
迈克尔·格雷夫斯, 1967-71,
轴测图, 体量图解;
Hanselmann House,
Michael Graves, 1967-71,
axonometric drawing, massing
diagram;

十字形教堂, 被置于一个大且深的梯形空腔中。

除了实际施工操作外, 加法与减法同样可以用来表达建造形式的感知特征。这里我们选取了两个建筑案例。其一是安德烈亚·帕拉第奥设计的圆厅别墅 (约1570), 位于意大利卡普拉洛拉郊外的一座大型古典住宅。从外部来看, 建筑中央是一个立方体块, 四个侧面对称设置附加门廊, 顶部有一个穹顶 [图2.2-5]。这些要素有着相对独立的识别特征, 整栋建筑可以想象为各部分的加法构成。在迈克尔·格雷夫斯设计的汉索曼住宅 (1967-71) 中, 主体的居住体量也是一个立方体 [图2.2-6]。在它的前方, 有一个分离的独立小建筑及片墙, 它们与主体量之间通过桥连接。主体量尺寸约为10.45m见方、8.75m高, 其南面角落被削去并用玻璃幕墙围合, 在上层部分引入了开放阳台。这一设计特征是减法的, 体现为立方体实体中存在的一个复杂虚空。

组织构思与场地

关键概念: 构思, 场地划分, 重心, 田字格, 九宫格。

建筑设计很少是关于简单的单一实体体量或虚空容积。建筑是复杂的, 它的形式与功能要求之间通常会发生冲突。另外还存在与场地、周围环境之间的关系。若要在一个具有结构与秩序的形式布局中, 解决这些相互对抗而独立的空间需求, 就需要建立一个系统: 一个组织策略或模式。在巴黎美院的建筑设计教学中, 法语单词"Parti"是用来区分要素组织模式设计构思的基本概念。通常可以用它描述平面上的空间构成, 所指的是三维空间上的配置。它也会被用来描述剖面上的空间组织, 如卡赛基住宅中相扣的双层通高空间, 或城市形态布局, 如古罗马军营采用的田字格格局(位于阿尔及利亚的提姆加德)。

场地所指的是一个具有明确界限划分的区域或表面, 其形状是可辨别的。在构成中, 一块场地可能会与其他的场地或物件重叠, 或局部重合。通过形状与比例, 可以确定场地的几何性质。比如说, 一个长短边比例为1:1.618 (黄金分割) 的长方形场地, 可以分割为一个正方形场地与另一个具有黄金分割比例关系的长方形场地。这样的场地划分并不是随意的, 因为它确实了该形状所具有的潜在几何属性。

空间中的场地同样具有视觉能量或动力 [图2.2-7], 并对场地内外的要素布置产生影响。场地的几何性质在很大程度上决定了力的模式。鲁道夫·阿恩海姆分析了方形的特征与性质。方形上的某些点相较其他点来说具有更强的视觉吸引力。在方形上, 最突出的点是它的中心, 即重心、视觉平衡

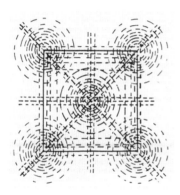

图 2.2-7
方形场地的视觉动力图解;
forces of a square field;

acter of built form rather than the actual construction process. Two houses will serve as examples. The first is the Villa Rotondo (ca. 1570) by Andrea Palladio, a large classical-style residence in the countryside of Caprarolla, Italy. Visually, the building is a cubic block with four identical porticos symmetrically attached and a dome projecting through the roof [fig. 2.2-5]. These elements have individual identity and the building can be imagined as an additive composition of parts. The main living block of the Hanselmann House (1967-71) by Michael Graves is also a cubic form [fig. 2.2-6]. In front, a separate smaller building and a detached façade screen wall are connected to the main building by a bridge. The main block measures approximately 10.45m x 10.45m x 8.75m and is eroded on the south-facing corner by setting the enclosing glass wall back and introducing an open balcony at the upper level. This design feature is subtractive and suggests a complex void in an otherwise solid cubic mass.

parti and field

Key concepts: *parti, field subdivision, center of gravity, four-square, nine-square.*

An architectural design is rarely a simple solid mass or an empty void. Buildings are complex objects with conflicting formal and functional requirements. In addition there is always a relationship to the site and surrounding context. To resolve these competing and interdependent program demands into a formal arrangement that has structure and order requires a system, an organizational strategy or scheme. The word *parti* is French and was used in the Ecoles des Beaux Arts to identify the basic concept of a scheme as a pattern or arrangement of elements. Typically the parti is a description of a spatial composition expressed in plan but referring to a three-dimensional configuration. It can also be used to describe an arrangement that's primarily sectional (interlocking double height spaces, a la Villa Carthage) or even an urban pattern (four square pattern of Roman Castrum, a la Timgad in Algeria).

A *field* is a term used to identify a demarcated area or surface. It is bounded and has a recognizable shape. A field may exist in a composition where it may be overlapping or partially coincident with other fields or objects. A field has geometric properties determined by its shape and proportions. For example, a rectangular field that has a ratio of its long to short sides of 1.618, a golden section

图 2.2-8
圆片的位置与方形形式的力的作用;
the position of a circular disc and the force of a square form;

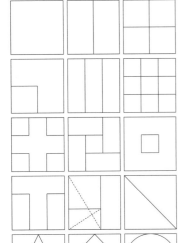

图 2.2-9
方形的几何性质与划分;
properties and subdivisions of a square;

点。连接方形的两条对角线会在中心位置相交，连接方形两对边中点的线亦然。从视觉重要性的角度来说，方形的四角仅次于它的中心。阿恩海姆阐释了要素摆放与这些基点之间关系的视觉影响，比如说将一个圆片放置于方形上的非中心位置就会表现出不稳定的状态 [图2.2-8]。当圆片往方形的边缘移动，会产生强烈的视觉张力将其"推离"边缘。

方形的几何属性暗示了其场地划分的不同可能 [图2.2-9]。其中，具有特别趣味的两种是田字格与九宫格。田字等分方形是最基本的一种网格形式：对边的正交等分线相交于一点并建立中心。不同于九宫格中网格将方形等分为9个小的正方形，且其中之一占据了中心位置，田字格在组织上是外围式的，中央是交叉形成的点：实体而非空间。这两种完全相对的情况，暗示了不同的组织策略或构思。田字格产生的是一个相对中性的配置，四个空间分别占据方形的四个角落，并没有哪个空间更为突出 [图2.2-10]。然而，九宫格创造了一个层级关系：中央的场地与原方形的中心重合，在强化中心性的同时也强调了这一位置的首要性 [图2.2-11]。

模式的构思组织在本质上可表现为图解。在不丢失其基本组织结构的同时，还可以发展出很多独特、不同的设计结果。16世纪帕拉第奥在意大利威尼托区域设计的一系列九宫格住宅就是经典案例之一。艺术史学家鲁道夫·维特科尔对此进行了研究，以表明文艺复兴时期帕拉第奥住宅的类型连贯性 [图2.2-12]。

模数与度量
关键概念：模数，均分，控制线，比例，公度比例，黄金分割比例，斐波那契数列，模度。

每个建筑都是不同部件的复杂组织。这些部件包括结构、围合、室内装饰、楼梯，甚至楼面地砖的实体构件。为使得这些部件可以在尺寸上协作，就需要建立一个系统，也就是模数系统。模数是度量的基本单位，它重复地贯穿应用于构成或建筑设计中。它可以是出于结构的考量，比如取柱间距或砌体构造单元。设计的其他组成部分就需要采用与这个模数或该模数整数倍相对应的尺寸，这样才能够实现对位与配合。

模数是一种强大的调节工具，通过模数的重复可以赋予构成以秩序。但是，一个模数系统并不完全排除变形，亦无需限制所有的尺寸以实现完全一致。其关键在于"均分"：我们可以将一个模数均分为二、三或更多部分，这样无论各部分的尺寸大小如何，它都能够通过复制达到模数尺寸。同样地，还可

图 2.2-10
以田字格划分为设计组织策略的学生习作；
student project with a parti based on four-square subdivision;

图 2.2-11
以九宫格划分为设计组织策略的学生习作；
student project with a parti based on nine-square subdivision;

图 2.2-12
帕拉第奥住宅平面分析图解，鲁道夫·维特科尔，1949（图中文字从左至右，由上及下依次为：蒂内别墅，齐科纳；萨莱哥别墅，米加；波亚纳别墅，大波亚纳；巴杜尔别墅，弗拉塔，波莱西内；泽诺别墅，切尔尼托；科尔纳别墅，皮翁比诺德塞；皮萨尼别墅，蒙塔尼亚纳；埃莫别墅，范佐洛；马尔孔腾塔别墅，米拉；皮萨尼别墅，巴尼奥洛；圆厅别墅，近维琴察；帕拉第奥别墅的几何模式）；
diagram analyses of Palladio's villa plans, Rudolf Wittkower, 1949;

rectangle, can be subdivided in such a way that a square field and another rectangular field of golden section proportion is produced. The subdivision of a field in this way is not arbitrary as it acknowledges the underlying geometric properties of the shape.

A field in space also contains visual energies or forces [fig. 2.2-7] that exert influence on the position of elements within and outside of the field. The geometry of the field largely determines the pattern of forces. Rudolph Arnheim has identified these characteristics or properties in the shape of a square. There are certain points or spots on the square that attract more visual attention than other spots. The center point of the square, which is also its center of gravity and point of visual balance, is the most dominant. The diagonals from the corners of the square intersect at the center, as do the perpendicular lines from the midpoints of each of the sides. The four corners of the square are next after the center in terms of visual importance. Arnheim demonstrates the visual effect that an element's placement has in relationship to these cardinal points. A circular disc, for example, will appear unstable when placed close to but not over the center of the square [fig. 2.2-8]. As it migrates towards the edge of the square a strong visual tension is generated that "pushes" the disc away from the edge.

The properties of a square suggest many possible subdivisions of the square field [fig. 2.2-9]. Two that are of special interest are the *four-square* and *nine-square*. A basic four-part subdivision of a square is a grid in its most reduced form. The intersecting grid lines create a point at their crossing, establishing a center. But unlike the nine-square which is a square subdivided by grid lines into nine rectangles or squares with one occupying the center, the four-square is peripheral in its organization and the center is a point of intersection, a solid not a space. These two opposing conditions suggest different organizational strategies or partis. The four square tends to produce a slightly more neutral configuration in which the four spaces each occupy a corner of the square field and none are exceptional. The nine square on the other hand, establishes a hierarchy in which the center square coincides with the center of the square field and reinforces the centrality and dominance of this position.

A parti is a schematic organization and essentially a *diagram*. It can develop into many unique and individual designs without losing its basic organizational structure. A classic example is the variation of the nine-square parti in the villa designs of Andrea Palladio

27

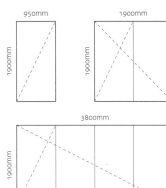

图 2.2-13
要素的复制及其比例关系；
repetition of element and
proportional relationship；

950mm

1900mm

1900mm

1900mm

3800mm

1900mm

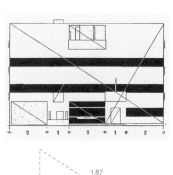

图 2.2-14
斯坦因别墅，
勒·柯布西耶，1927，
立面比例分析图解；
Villa Stein, Le Corbusier, 1927,
proportional analysis of the
façade；

1

1.87

1.59

以通过复制单位模数，即将模数乘以整数倍来获得更大尺寸。如果是整数倍，我们可以获得2:1、3:1等倍数关系；如果是整数比，则可以是1/2、1/3等分数关系。因此，模数系统就允许一个模数尺寸的要素与其他具有相同尺寸或成倍模数尺寸的要素相配合。

板片构件（如门窗）有长、宽两个主要尺寸。对于矩形要素，其比例可以由两侧边的比值表示。比如说一道高1900mm、宽950mm的门，它的比例为2:1。将两道同样尺寸的门并置在一起，就形成了1900mm x 1900mm的方形；四道门并置则是双正方形，尺寸为3800mm x 1900m，但其比例同单扇门一样 [图2.2-13]。

通常会绘制矩形的对角线以辨别，表示其比例关系。任何两个相同比例的矩形，其对角线也会互相平行。如果将一个矩形分割为两个矩形，其中一个矩形的对角线与原矩形的对角线相垂直，那么这两个矩形的比例相同。这一技巧适用于探索、创造比例关系，通常称之为"控制线"。其中的一个例子就是对勒·柯布西耶设计的斯坦因别墅的立面分析，通过绘制控制线以表示正立面上窗、门、阳台等要素的比例相同。值得注意的是，其立面外轮廓也采用了同样的比例 [图2.2-14]。

如果两边的比值可以用整数表达（如1:2；3:4等），我们称之为"公度比例"。建筑师帕拉第奥在规划弗斯卡利别墅的房间尺寸时，使用了一组简单的公度比例：3:4，4:4和4:6。这些比例所对应的房间尺寸分别是12'x16'（约3.67m x 4.88m，3:4），16'x16'（约4.88m x 4.88m，4:4）以及16'x24'（约4.88m x 7.32m，4:6）。十字形拱顶空间的尺寸则取决于这些房间的位置以及网格线的延伸 [图2.2-15]。帕拉第奥及其他文艺复兴时期的艺术家重新探索了最早由毕达哥拉斯（公元前6世纪）发现的关于音程（三度、五度、八度等）与数值比值之间关系的推论性原则。比如说当两个音弦的长度是2:3，那么每个音弦的音高差别就是五度。同理，长度比值为3:4时，则产生四度音程。对于帕拉第奥而言，这就意味着普适和谐的存在，它以整数比值为基础，可以用来创造建筑与音乐中的美感。

历史上还有一些比例关系被认为是迷人、通用的。其中最重要的一个就是"黄金分割"。欧几里得（公元前约325年 – 约265年）最先定义了该比例并进行了几何推导：取正方形的一半的对角线，将其旋转至方形的底边来确定新的长方形的底边边长 [图2.2-16]。如果将这个长方形的短边（方形的另一侧）定义为1，那么长方形的长边就是1.618，两边比值1:1.618就是黄金分割比例（0.618）。这一特殊比例的强大性质体现

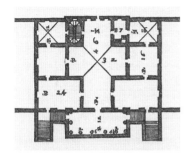

图 2.2-15
弗斯卡利别墅平面，
安德烈亚·帕拉第奥，图解；
plan of Villa Foscari,
Andrea Palladio, diagram;

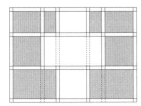

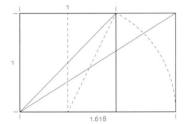

图 2.2-16
黄金分割比例的几何推导；
geometric derivation of the
Golden Section;

in the region of the Veneto of Italy in the 16th century. This was identified by the art historian Rudolf Wittkower and reflects a typological consistency in Palladio's villa of this period in Renaissance architecture [fig. 2.2-12].

module and measure

Key concepts: *module, subdivision, regulating line, proportion, commensurate proportions, Golden Section ratio, Fibonacci Series, Modulor.*

Every building is a complex organization of many parts. These parts are real physical components of structure, enclosure, interior finishes, stairs, and even floor tiles. In order for all of these parts to work together dimensionally, a system is required. One such system is *modularity*. A module is a unit of measurement that is repeated throughout a composition or building design. It may be a structural condition such as the spacing between columns or a masonry construction unit. Other components of the design will have dimensions that match the module or some multiple of the module's dimension. In this way there will be alignment and compatibility.

The module is a very powerful regulating device. It imposes order on a composition by its repetition. But a modular system does not necessarily exclude variation nor limit all dimensions to being identical. The key is in *subdivision*. A module can be subdivided into two, three or any equal number of parts and whatever the dimension of these parts is, it will fit the module in some multiple. Likewise a dimension can be obtained by multiplying the module by some number. If it is a whole number we obtain ratios like 2:1, 3:1 and so forth. These are referred to as whole number ratios and they relate to fractions such as 1/2, 1/3, etc. The system of modularity therefore allows an element with a modular dimension to match other elements with the same dimension or a multiple of the dimension.

Planar elements, a door or window for example, have two principle dimensions, length and width. Being rectangular, the proportion of the element is described by the ratio of the two sides. So for example, a door that is 1900mm in height and 950mm in width has a proportion of 2:1. Two doors of the same proportion can align side by side forming a rectangle of 1900mm X 1900mm, a square. Four doors aligned form a double square, 3800mm x 1900mm that has the same proportion as a single door [fig. 2.2-13].

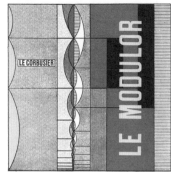

图 2.2-17

《模度》，勒·柯布西耶;
The Modulor, Le Corbusier;

在它的分割上：如果我们将一个黄金矩形分割为两个矩形，其中一个是正方形，那么另一个长方形同样具备黄金分割比例。

来自于几何的成比例比值常表现为无理数（不可用两整数之比表示），比如说 π 就是一个无理数。这些比例则被称作非公度比例，黄金分割比例就是其中一种。然而，意大利数学家斐波那契发现黄金分割比例有另一种非常特别的性质。一组以1开始的数列，第二位是2，其后的每个数字都是前两位数字相加的结果。该数列的前十位是：1, 2, 3, 5, 8, 13, 21, 34, 55, 89。在这个数列中，两连续数字间的比例在数值上会逐渐趋于黄金分割比例，比如说55与89，两者的比值是0.6179。就算是位于数列前位的5与8，其比值0.625之于黄金分割比例的差别也在1%以内。勒·柯布西耶意识到，可以从斐波那契数列整数的比例系统出发，创造出接近于黄金分割的比值。他基于这一想法研究创造了一个极具创意且重要的比例系统：《模度》[图2.2-17]。关于模数，阿尔伯特·爱因斯坦曾总结："正是有了比例的尺度系统，使得做好不甚容易，做差也有难度。"[1]

1   Le Corbusier, *The Modulor: A Harmonious Measure to the Human Scale, Universally Applicable to Architecture and Mechanics* (Basel; Boston: Birkhäuser, 2000): 58

The diagonal line drawn between opposite corners of a rectangle is a visual convention that relates to and helps identify proportional relationships. Any two rectangles of similar proportion will have parallel diagonal lines. If we subdivide a rectangle into two rectangles, and the diagonal of one of the new rectangles intersects the original at 90°, then the new rectangle is of the same proportion as the original. This technique is useful in discovering and creating proportional relationships. It is referred to as the *regulating line*. The analysis of the façade of the Villa Stein by Le Corbusier is an example of the use of the regulating line to illustrate the similar proportions of elements such as windows, a door, a balcony, etc. that compose the front elevation. Note that the elevation as a whole is also of the same proportion [fig. 2.2-14].

If the ratio between two sides is expressed in whole numbers (1:2, 3:4 and so forth) we call these commensurate proportions. In planning the dimensions of the rooms in the Villa Foscari, the architect Palladio used simple commensurate proportions: 3:4, 4:4 and 4:6. These ratios correspond to room dimensions of 12' x 16' (3:4), 16' x 16' (4:4) and 16' x 24' (4:6). The dimensions of the cruciform shaped main vaulted space result from the position of these rooms and the extension of the grid lines [fig. 2.2-15]. Palladio and other renaissance artists rediscovered the principle first observed by Pythagoras (6th century B.C.) of a corollary between musical intervals (third, fifth, octave, etc.) and number ratios. For example, when the lengths of two musical strings are in relation to each other by 2:3, then the difference in the pitch of the sound of each string is a fifth musically. In the same manner, the ratio of 3:4 produces a fourth. For Palladio it meant that there was a universal harmony that produced beauty in both architecture and music. It was based on whole number ratios.

There have been other proportional ratios historically that were also considered beautiful and universal. The most important one is known as the *Golden Section*. Euclid (c. 325 – c. 265 B.C.) first defined the proportional ratio and gave its geometric derivation: the length of the half diagonal of the square rotated to the base of the square defines the length of a rectangle [fig. 2.2-16]. If the short side of the rectangle is defined as 1 (the side of the square), then the long side of the rectangle is 1.618, and the ratio of 1:1.618 is the Golden Section ratio (0.618). The most powerful property of this particular proportion is in its subdivision. If we divide a golden rectangle into two rectangles, one of which is a square, then the other rectangle

will also have the proportion of the golden section.

Proportional ratios derived from geometry often result in irrational numbers, numbers that cannot be expressed by ratios of integers. The ratio π for example is an irrational number. These ratios are referred to as incommensurate proportions. The Golden Section ratio is an incommensurate ratio. However, Fibonacci (1170 -1250) the Italian mathematician, discovered that the Golden Section ratio had another remarkable property. That the ratio between successive numbers in a particular numerical sequence of whole numbers beginning with 1, followed by 2 and then followed by the number that is the sum of the previous two numbers, converges on the Golden Section ratio. The sequence in its first eight numbers would be: 1, 2, 3, 5, 8, 13, 21, 34, 55, and 89. If we take the last two, 55 and 89, their ratio is 0.6179! In fact even the ratio between 5 and 8 is 0.625: within 1% of the Golden Section ratio. Le Corbusier realized that a proportional system based on the whole numbers of the Fibonacci series would produce ratios that approximated the Golden Section. He based the invention of *The Modulor* [fig. 2.2-17], one of the most innovative and important proportioning system ever devised, on this concept. About the Modulor, none other than Albert Einstein decalred: "It is a scale of proportions which makes the bad difficult and the good easy." [1]

2.3

构成［空间］
composition [space]

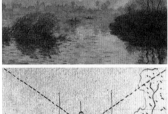

图 2.3-1
《拉维科特的塞纳河日落，冬天景色》，克洛德·莫奈，1880；
Soleil couchant sur la Seine à Lavacourt, effet d'hiver,
Claude Monet, 1880;

空间限定
关键概念：空间作为一种具有形状、尺寸与尺度的体积形式；以点、线、面限定空间；空间之间的运动（流线）。

　　人们对于空间的认识存在一个固有观念：空间是无边无际的、开放的，可随视线无限延展。它与远处的地平线或外太空中无穷尽的黑暗共同组成了一个开放的领域。地平线是普遍存在的，它是天空与地面之间所共享的一条视觉边界。当人们站在月球或其他的行星上，凝视着无限空间中的黑暗，地平线依然存在。

　　在绘画艺术中，艺术家发展了不同的方法以在二维的画布上再现三维空间。其中一种涉及画面中前景、中景与背景的布置：三者分别对应表现距离观察者由近及远的空间，并分配于画面的下、中、上三部分区域，如克洛德·莫奈的作品《拉维科特的塞纳河日落，冬天景色》（1880）[图2.3-1]。儿童在绘画时会本能地以类似方式表达空间距离，并理解"近大远小"的视觉效果。15世纪时，艺术家发现了透视的再现方法，这使得对于空间深度的真实描绘成为可能。建筑师在此基础上展开思考，并想象其在建筑中的应用。

　　建筑空间，主要是一个内部整体，由房间的四面墙、楼面与顶棚包裹容纳。这一空间的限定是明确的：房间的围合界面明确决定了其形态、尺寸。将两个这样的房间并置排布在一起，并在其间创造一个开口（门洞或通道），这样两个空间便彼此连通，人可以从一个房间穿行到另一个房间中。当很多房间通过一组相互对位的门洞联系在一起时，就会产生一条连续的、满足流线需求的通道，我们称之为纵贯。亦可以通过独立的要素（如通道或走廊）来连接这些房间，这些空间要素与人的运动直接相关。在比例上，这些要素的深度长过其宽度。要素可沿着房间排布，或置于房间之间，穿插于房间之中。

space definition

Key concepts: *space as a volumetric form possessing shape, dimension and scale; definition of space with point, line and plane; movement between spaces (circulation).*

The concept of *space* that is innate to humans is that of an unbounded, open space extending as far as the eye can see. It is an open terrain with a distant horizon or the infinite blackness of outer space. The horizon is always present and is the visual shared contour line between the sky (heaven) and the ground (earth). A horizon also exists on the moon or any other planetary body that a man might stand upon and gaze into the blackness of deep space.

In painting, artists have developed conventions to represent the third dimension of space on a two-dimensional flat canvas. An example involves assigning the foreground, the space closest to the viewer, to the lower portion of the painting; the middle ground to the middle and the background, which is the most distant, to the upper portion. In the Claude Monet painting "Soleil couchant sur la Seine à Lavacourt, effet d'hiver" (1880) [fig. 2.3-1], the water and two islands are the foreground, the silhouette of the village sits on the horizon or middle ground, and the sky is the background. Children instinctively incorporate spatial distance in this way into their drawings, but not incorporating size diminution with distance. In the 15th century, artists discovered the representational technique of *perspective* that enabled a realistic depiction of spatial depth. Architects pondered the implications of this new vision and imagined its application in architecture.

Architectural space is primarily an interior entity and originally was bounded and contained by the four walls, floor and ceiling of a room. The definition of such a space is explicit: the surfaces of the enclosing planes of the room unambiguously determine the configuration and dimensions of such a room. Place two such rooms side by side and create an opening between them (a doorway or portal) and the spaces are connected; a person can move from one room to the next. Many rooms so connected with their doorways in alignment create a continuous path for circulation that we call *enfilade*. Alternatively the rooms may be connected by a separate element, a passageway or corridor that is a space designed primarily for movement. Its dominant proportion is long with respect to its width. The position of this element to the rooms can be alongside, between or

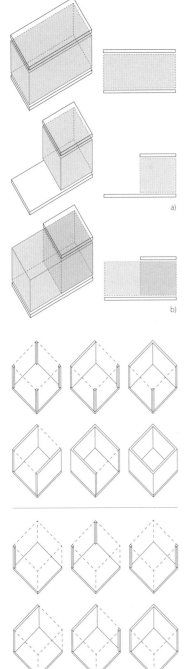

图 2.3-2
空间限定与空间阅读；
space definition and its reading；

a)

b)

图 2.3-3
利用板片与柱子来创造明确的空间限定（上）与暗示的空间限定（下）；
explicit (up) and implicit (below) spatial definitions with planes and columns；

那么，最少需要多少要素才可以限定一个空间？如果我们取走限定房间的一片或多片板片（墙体、楼面或顶棚），这个空间是否依旧被限定？换句话说，我们是否还能够准确感知该空间的形式？如果将所有的墙体都移走，情况会有什么变化？现在，我们仅保留悬浮的顶棚板与楼面，保证其位置不变且各角落完全对齐 [图2.3-2]，我们是否能感知原来的空间？如果再进一步，移走全部或局部的顶棚板，情况又如何？我们发现原本空间体量的边界开始变得模棱两可：究竟是将顶棚板边角投形到楼面上所囊括的空间体量，还是将楼板的边角投形到顶棚板所在位置上产生的空间体量？

通过这样的讨论，我们可以得到一个结论：为保持原长方体体量边界的清晰可读，所需要的最少限定是空间的八个角点，本质上就是立方体形式的各角落。不论如何组合限定空间的边缘、角落或界面，只要明确了这八个角点的位置，就能对空间体量的形式作清楚限定 [图2.3-3]。

这一原则在现代建筑中的意义是革命性的。受益于建造技术，包括使用楼层通高的玻璃与长跨度的结构悬挑，在限定空间体量形式的同时，还能够呈现出无边界的视觉效果。现代建筑师在其实践中均有在探索这种潜力，如密斯·凡·德罗的巴塞罗那展览馆（1927）[图2.3-4]。

空间的重叠与现象的透明

关键概念：空间的重叠，相互渗透，物理的透明，现象的透明，立体主义，空间的层。

直到20世纪早期，建筑主要是由一组房间组合构成，各空间清楚限定并由承重的厚墙分隔。伴随着19世纪末钢铁与钢筋混凝土结构框架的出现，室内空间的性质开始发生改变。墙体不再是唯一的结构支承要素，相反，柱作为支承结构成为更加经济的一个选择。这使得房间的空间限定可以更加自由，墙体相应转变为非承重的空间隔断，可以单纯考虑建筑意图来对其操作。

建筑师弗兰克·赖特（1867-1959）在其早期建筑作品布劳萨姆住宅（1892）中，探索了平面上的空间开放性：通过大的拱形开口将房间联系在一起，拱券与室内的其他主要要素相对齐 [图2.3-5]。虽然布劳萨姆住宅在风格上带有折中与古典倾向，但是在室内与室外的种种细节中，我们可以看到其对于"水平带"的表达，这预示了赖特草原式住宅的一个典型特征——延展的水平性。

在1904年的马丁住宅设计中，赖特对建筑空间的限定做

图 2.3-4
巴塞罗那展览馆，
密斯·凡·德罗，1927；
Barcelona Pavilion,
Mies van der Rohe, 1927;

图 2.3-5
布劳萨姆住宅室内，
弗兰克·赖特，1892，
平面空间图解；
Blossom House (interior),
Frank Lloyd Wright, 1892,
spatial diagram in plan;

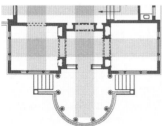

in-between.

What is the minimum definition of a space? If we remove one or more planes (walls, floor or ceiling) that define a room, is that space still defined? In other words, are we able to visualize the precise form of the space? What if all the walls are removed? Now we have a condition of a floating ceiling plane and a floor. These two planes remain in their original position with corners that align perfectly. [fig. 2.3-2] Can we still visualize the original space? What if we go further and remove all or part of the ceiling? We discover that the boundaries of the original spatial volume are becoming slightly ambiguous. Is the space that volumetric form that can be visualized by projecting the corners of the remaining ceiling plane down to the floor? Or is it the original room space whose boundary can be determined by projecting the edges and corners of the floor plane vertically to the height of the floating ceiling plane?

Through such an investigation we reach a conclusion that the minimum requirement for the precise determination of the boundaries of the original rectangular volume are eight points in space, essentially the corners of the cuboid form. Any combination of edges, corners or planes that can mark the position of these eight points in space thus provides a precise determination of the form of the spatial volume [fig. 2.3-3].

The importance of this principle in modern architecture was revolutionary. Aided by the introduction of new building technology such as floor to ceiling glass and long structural cantilevers, space could be defined as volumetric form and appear boundless at the same time. Modern architects such as Mies van der Rohe exploited this potential in his critical works such as the Barcelona Pavilion (1927) [fig. 2.3-4].

spatial overlap and phenomenal transparency

Key Concepts: *spatial overlap, interpenetration, literal transparency, phenomenal transparency, cubism, spatial layering.*

Until the early twentieth century, buildings were primarily composed of rooms, discrete well-defined spaces separated from each other by thick walls that were typically the load bearing structure. With the emergence of the steel and reinforced concrete structural frame in the late nineteenth century, the nature of interior space began to change. No longer were walls required for structural

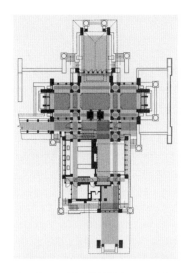

图 2.3-6
马丁住宅,
弗兰克·赖特, 1904,
平面空间图解;
Martin House,
Frank Lloyd Wright, 1904,
spatial diagram in plan;

图 2.3-7
至圣救世主教堂内部,
安德烈亚·帕拉第奥, 1592,
平面空间图解;
Church of the Most Holy
Redeemer (interior),
Andrea Palladio, 1592,
spatial diagram in plan;

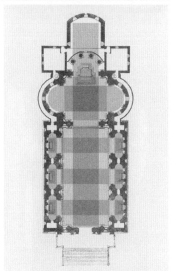

出了进一步提升。在建筑中使用梁与柱作为主要的承重结构,并将其结构支撑功能与要素的空间限定作用相结合,创造出复杂的空间限定与重叠[图2.3-6],这一点影响了现代建筑的发展。从外部来看,通过引入长跨度的屋顶悬挑、平开窗的镶边、窗台与过梁的对位以强调水平性;对于内部空间,露明的砖柱砌缝、木梁、连续同高的窗台与过梁,以类似的方式强化了水平线条。通过巧妙地布置结构支撑及其尺寸,以及重叠的空间区域,创造出了动态但是清晰可读的复杂空间关系。

在过去的一些建筑作品中,如安德烈亚·帕拉第奥在威尼斯设计的至圣救世主教堂(威尼斯救世主,1592),同样暗示了独立限定的空间的复杂构图:空间彼此重叠且互相渗透[图2.3-7]。在设计该教堂时,如何协调传统的教堂中殿或巴西利卡平面形制与更为古典的集中式平面是帕拉第奥遇到的主要挑战。为实现这一想法,帕拉第奥使用了重复的壁柱母题将两个主要的空间融合在一起,既联系各独立空间的同时又明确了空间的重叠部分。

通过使用结构框架与玻璃幕墙(如包豪斯校舍[图2.3-8],1926)而实现的物理性透明,开始成为现代建筑的一大特征。其后,在一篇名为《透明性:真实的与现象的》[1]的重要论文中,作者提出了一个关于空间透明性的论断:当两个或多个清晰限定的空间体量或区域彼此相重叠与融合[图2.3-9],但同时各空间仍具有可读性时产生的空间透明被称作是现象的透明。它取决于空间组织的品质,而非材料(如玻璃)的物质属性。在这篇文章中,作者还指明了立体主义艺术运动中与此平行的类似倾向,例如是以乔治·布拉克、巴勃罗·毕加索与胡安·格里斯为代表的综合立体主义[图2.3-10]。

将"现象的透明"作为一种复杂却具有视觉秩序的空间构成技巧,在勒·柯布西耶的作品中得到了呈现。以斯坦因别墅(1927)为例,在其二层平面中,我们可以辨识出一个比例为1:2:1:2:1的结构网格,其上巧妙排布与塑形的分隔墙,连同条形窗及垂直洞口一起,创造出了空间区域间的重叠,这是对现象透明性的典型展示[图2.3-11]。

图 2.3-8
德绍包豪斯校舍，
瓦尔特·格罗皮乌斯，1926，
物理性透明；
Bauhaus Dessau,
Walter Gropius, 1926,
literal transparency;

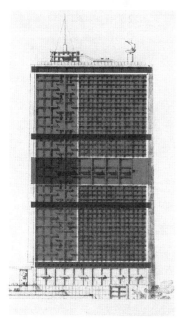

图 2.3-9
阿尔及尔塔楼，
勒·柯布西耶，1939，
现象的透明；
Algiers Tower, Le Corbusier,
1939, phenomenal
transparency;

图 2.3-10
葡萄，胡安·格里斯，1920；
Grapes, Juan Gris, 1916;

support as columns could provide the required support more economically. As a result, the spatial enclosure of rooms could be more flexible and the wall became essentially a non-loadbearing partition, easily manipulated for purely architectural consideration.

The architect Frank Lloyd Wright (1867-1959), in one of his early works, the Blossom House (1892), explored spatial openness in plan, connecting rooms through large arched openings with interior features in alignment [fig. 2.3-5]. Although eclectic and classical in style, the Blossom House has a horizontal banding expressed in finish detailing both on the exterior as well as the interior that presages the extensive horizontality that became a key characteristic of the later Prairie House style.

In the Martin House of 1904, Frank Lloyd Wright established a new level of refinement in the development of space definition. Using piers and columns as the principle load bearing structures, Wright coordinated the structural support function with the space defining role of these elements to create a complexity of space definition and overlap [fig. 2.3-6] that would influence modern architecture. On the exterior, horizontality is enhanced by the introduction of long roof cantilevers, bands of casement windows and the alignment of sills and lintels. On the interior, spaces are similar banded by horizontal lines created by the exposed brickwork of the piers, the wood trimmed beams and the continuous height of the windowsills and lintels. Space is carefully articulated by the size and placement of structural supports, and the overlap of spatially defined zones creates a complexity of space that is both dynamic and legible.

Some works of the past, such as the Church of the Most Holy Redeemer (Il Redentore, 1592) in Venice by Andrea Palladio, suggest a complex spatial composition of individually defined spaces that overlap and flow into other spaces [fig. 2.3-7]. Palladio's challenge in designing the church was to reconcile the traditional nave or basilica plan type with the more classical centralized plan. To achieve this, Palladio had to merge the two dominant spaces together, which he accomplished through the use of a repetitive attached column motif that articulated the individual spaces while also identifying the overlap spaces.

With the introduction of structural frames and glass curtain walls (e.g. Bauhaus, 1926 [fig. 2.3-8]) a type of literal transparency emerged as a characteristic of Modern Architecture. Later, in an important essay entitled "Transparency: Literal and Phenomenal"[1],

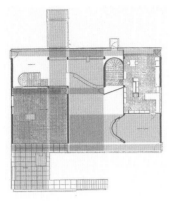

図 2.3-11
斯坦因别墅,
勒·柯布西耶, 1927,
平面空间图解, 网格图解;
Villa Stein, Le Corbusier, 1927,
spatial diagram in plan, grid
diagram;

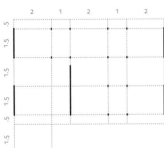

建筑的原理与要素

关键概念: 秩序, 表现力, 复杂性, 和谐, 轴线, 对称, 层级, 基准, 重复, 方向性, 正面性, 集中的, 外围的, 对边对称与四边对称, 局部对称, 显著尺寸, 串联。

如同其他的视觉艺术, 建筑中概念的表现与传达可以通过构成来实现, 也就是对设计要素的选择、排布与配置。构成中一个概念的连贯表达, 既不模棱两可也不自相矛盾, 依赖于秩序的建构。

"当秩序被认为是可有可无的, 或是可以由其他事物替代的一种品质时, 最后的结果只会是混乱。我们需要明白, 秩序是任何有组织性的系统, 发挥其(物理上或心理上的)效用所不可或缺的。不论是机器, 管弦乐队还是球队, 缺乏要素、成员间的综合协作, 它都无法正常运转。这对艺术或建筑来说是相同的, 它只有在呈现为一个有序的模式时, 才能满足功用并传达讯息。秩序可以在不同的复杂程度上实现: 简单的如复活节岛上的石像或一座农舍, 复杂如贝尔尼尼的画作、波罗米尼的教堂。但是如果没有秩序, 就无法述说一个作品想要传递的信息。"[2]

鲁道夫·阿恩海姆的这段文字着重陈述秩序与模式(或者说有组织的结构)之间的关系。在建筑中, 功能会对组成建筑设计中的诸多方面造成影响。具有不同使用要求及其对应面积的空间分配, 即空间内容计划, 是其中的一个主要限制。另外, 重力会对与建筑结构相关的跨度、力的分配与传递路径产生限制。需准确配置机械设备与建筑围护以控制气候造成的影响, 包括寒、热、降雨、湿度、日照与风。

建筑的原则与要素是指设计中建构秩序的概念或手段。轴线、对称、层级、基准与重复是构成中创造秩序的基本形式组织手法。不论是什么历史时期或风格的建筑, 均会呈现其中的一种或多种原则。形式分析的目的之一, 就是厘清这些原则是如何应用及其组合的。

轴线是一条假想的、无限长的线, 在这条线的两侧是对称布置的要素。它也可以是一条长度有限的线, 端点为空间中的两点。轴线的端点可以有不同的限定方式, 如一个垂直延伸的点(如独立支承的柱、方尖碑), 一个清楚限定的空间(如城市广场), 一个垂直板片(如建筑立面), 或是一个门洞、通道。最后一种情况, 轴线可能会穿过门洞向外延伸。

"轴对称性"与"正面性"是相关的两个概念。空间中的两点可以确定一条线, 将其垂直向上延伸可形成面。空间中的面有两面, 其中的一面可能会相对更受重视, 如墙体的外表面或建筑立面。可以由建筑立面确认一条线(轴线), 垂直于该

an argument was made for a condition of spatial transparency that exists when two or more well-defined spatial volumes or fields overlap and merge while simultaneously retaining their legibility [fig. 2.3-9]. This formal condition was called *phenomenal transparency* and relies on a system of spatial organization rather than a quality of material substance as with glass. In the same article, the authors identified similar parallel tendencies in the art of the cubist movement, particularly in the synthetic cubism of Georges Braque, Pablo Picasso and Juan Gris [fig. 2.3-10].

The work of Le Corbusier demonstrates the possibilities of phenomenal transparency as a technique of spatial composition that attains complexity with visual order. An example can be found in the first floor plan of the Villa Stein at Garches (1927). The structural grid with a 1:2:1:2:1 proportional zoning, strategic placement and shaping of partition walls, together with the strip windows and vertical slot openings produces an overlay of spatial zones that exemplifies the condition of phenomenal transparency [fig. 2.3-11].

principles and elements of architecture

Key Concepts: *order, expression, complexity, harmony, axis, symmetry, hierarchy, datum, repetition, directionality, frontality, centralized, peripheral, bi-lateral versus quadrilateral symmetry, local symmetry, significant dimension, concatenation.*

In Architecture, as in any of the visual arts, expression and communication of an idea is achieved through composition, that is, the selection, arrangement and configuration of the elements of the design. In order to express an idea coherently, without ambiguity or contradiction, a composition depends on order.

"Nothing but confusion can result when order is considered a quality that can equally well be accepted or abandoned, something that can be forgone and replaced by something else. Order must be understood as indispensable to the functioning of any organized system, whether its function be physical or mental. Just as neither an engine nor an orchestra nor a sports team can perform without the integrated cooperation of all its parts, so a work of art or architecture cannot fulfill its function and transmit its message unless it presents an ordered pattern. Order is possible at any level of complexity: in statues as simple as those on Easter Island or as intricate as those by Bernini, in a farmhouse and in a Borromini church. But if there is not order, there is no way of telling what the work is trying to say."[2]

Rudolf Arnheim's statement draws attention to the relationship between *order* and *expression* or, in other words, an organized structure. In Architecture, function places constraints on many of the elements that make up a building design. Program, the assignment of spaces with different uses and area requirements is a major constraint. Gravity imposes another set of limitations related to the structural requirements of span, load distribution and load path. Mechanical services as well as the building envelope must be configured correctly in order to control the effects of climate (heat, cold, rain, humidity, sun light, and wind).

The principles and elements of architecture refer to concepts or devices that create order in a design. These principles: axis, symmetry, hierarchy, datum and repetition, are the basic formal structuring devices by which a composition achieves order. Architecture of any historical period or style exhibits one or more of these principles. How they are used and in what combination is one

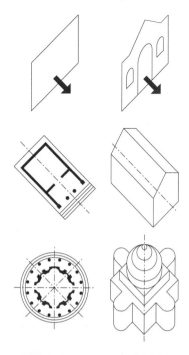

表面平面并穿越其中心或其他重要位置（如入口中心）。这一轴线可以用矢量线表示，穿越墙面并指向远离墙面的方向。矢量的这一方向性便确立了"正面性"（立面一词所表达的不只是"脸面"），并暗示了该建筑或物体的朝向。

美加仑（正厅）是指古希腊建筑中长方形的大厅或庙宇，其单一短边方向开敞并布置柱列。这种建筑基本单元有一个立面、入口及朝向。它是之后罗马神庙与基督教堂中采用的巴西利卡形式的原型。与美加仑相对的是圆形，最为理想的一种形状：圆上各点距空间中心的距离相同。圆形本身不存在任何朝向上的区别，但有很多轴线与其圆周切面相垂直。由圆形产生的集中式形式常被应用于宗教建筑中，并作为理想形式的宣言深受建筑师欢迎。以上这两种形式类型间的对立与终极抉择是早期建筑史上重要的争论之一。[图2.3-12]

在轴线两侧镜像布置要素便是创造"对称"，这是五点秩序原则中的第二点。对称在自然界中非常普遍，从模式上可以分为对边对称、四边对称与中心放射对称。若要在构成中对所有的要素进行对称布置，表现为镜面反射模式，通常会与设计中场地、使用计划、结构及其他功能因素相矛盾。出于这一考虑，大部分建筑设计会采用一个整体非对称的布局，而在其中可能存在一些局部对称。以住宅为例，它通常包含奇数量的独立空间需求，如起居室、厨房、餐厅及数间卧室等。若要在一个对称的平面中组织这些要素，并非不可能但充满挑战。古典主义建筑师托马斯·杰弗森在设计蒙蒂塞洛私宅时就遇到了这一矛盾，最后凭借其巧思创造了一个可接受的方案[图2.3-13]。

受制于现实情况（如不规则场地），局部对称可有效解决平面组织中的矛盾。马提尼翁酒店[图2.3-14]便是很好的一个案例。原为18世纪巴黎的联排别墅，这座建筑面临着一个困境：在不规则的场地中，入口处对称形式的庭院、立面与后方的规则式花园位于不同的轴线上。最终的解决办法是将两组对称布置的房间（局部对称）彼此重叠，使用对称轴重置的轴线将它们轻松地连接在一起。

在构成中，"层级"是通过建立等级与特殊性的差异来创造对比与秩序。缺乏层级性的构成可能是有序的，但是常会变得重复而单调。棋盘就是这样的一个例子：将一组大小相同、缺乏层级的方形以一种均质而重复的方式排布。作为现代写字楼建筑普遍特点的玻璃幕墙是另一典型：虽具有强烈的秩序感，但由于缺乏层级结构关系而显得平淡。在一些案例中，玻璃幕墙作为一个均质无特色的连续体被置于背景中作基准，而凸显前景中其他要素的独特与突出，如勒·柯布西耶于1932年设计的巴黎救世军庇护所[图2.3-15]。

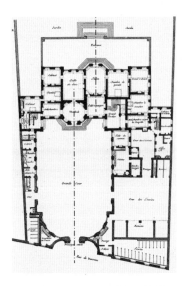

图 2.3-14
马提尼翁酒店首层平面，
简·库尔托纳, 1722-24;
Hotel Matignon, ground floor
plan, Jean Courtonne, 1722-24;

图 2.3-15
巴黎救世军庇护所，
勒·柯布西耶, 1932,
模型, 立面;
Salvation Army Building,
Le Corbusier, 1932,
model, elevation;

of the goals of *formal analysis*.

Axis is an imaginary line of infinite length implied by the symmetrical arrangement of elements on either side. It can also be a line of finite length established by two points in space that mark the endpoints. The endpoints of an axis can be defined in several ways: a point extruded vertically such as a freestanding column (obelisk); a well-defined space (urban plaza); a vertical plane (building façade) or a portal or gateway. In the last example, the axis may be extended through the portal and beyond.

Axiality is related to the concept of *frontality*. Two points in space define a line between them that when extruded vertically becomes a plane. A plane in space has two sides however one side might be favored, such as the exterior face of a building wall or façade. The façade plane implies a perpendicular line (an axis) positioned at the center of the wall plane or at a point of some significance (the center of the entrance). This axial line can be visualized as a vector both intersecting the wall plane and directed away from it. This directionality of the vector away from the wall plane establishes frontality (façade > "face") and indicates the orientation of the building or object.

The megaron is an ancient Greek rectangular hall or temple with one of the short sides open and framed by columns. This primordial unit of architecture has a façade, entrance and orientation. It is the origin of the *basilica* form that was later adopted by Roman temples and Christian churches. In contrast to the megaron is the circle, the most ideal platonic shape, created by a line equidistant from and surrounding a single point in space. The circle has no preference of orientation and yet has many axes perpendicular to tangential planes on its circumference. Favored by architects as a manifestation of ideal form, the circle led to the *centralized* form that was also adopted for sacred architecture. The opposition and ultimate resolution of these two formal types is one of the early and important theoretical debates in architectural history. [fig. 2.3-12]

The mirroring of elements on either side of an axis creates *symmetry*, the second of the five ordering principles. Symmetry is abundant in nature and can be bi-lateral, quadri-lateral or radial in pattern. Overall symmetry subjects every element of a composition to a reflective mirroring pattern that is often at odds with site, program, structure and other functional parameters of a design. For this reason, most building designs adopt an overall *asymmetrical* 41

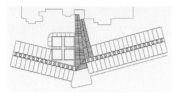

图 2.3-16
尺寸/形状的层级关系，奥利
维蒂培训学校，詹姆斯·斯特
林，1972，平面；
size/shape hierarchy, Olivetti
Training School, James Stirling,
1972, plan;

图 2.3-17
显著尺寸的概念；
the concept of significant
dimension;

图 2.3-18
费尔芒特自来水厂，
本杰明·拉特罗布，1872；
Fairmount Water Works,
Benjamin Latrobe, 1872;

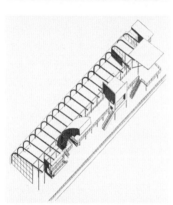

图 2.3-19
斯德哥尔摩展览会交通运输
馆，冈纳·阿斯普朗德，1930，
轴测图；
Stockholm Exhibition
Transportation Pavilion,
Gunnar Asplund, 1930, axon;

图 2.3-20
美国驻伊拉克大使馆住宅，
路易斯·舍特，1955-69；
United States Embassy
residence in Baghdad, Iraq,
José Luis Sert, 1955-69;

可以创造层级关系的方法很多。尺寸是一个明显特征，形状、位置及细节都可以用以创造差异，进而产生层级关系[图2.3-16]。问题的关键在于这些差异是否足够明确。正如概念"显著尺寸"所指，尺寸上的轻微递增或形状上的微小变化会被感知为误差或是错位，而最终使得对比与层级差无效[图2.3-17]。模棱两可的比例关系同样会使得构成预期意图表达的失败。

"基准"是五点秩序原则中的第四点。基准可以以不同的形式存在，它可以是线性的、面状的或体量的。基准是构成中为不同要素提供普遍参照的结构[图2.3-18]。它所创造的秩序可能会表现为不相干物件间的随机排布，其随机性从"模式不可识别"到"可辨形式的重复与高度组织"不等。不论是哪种情况，基准都可以通过提供一个不变的、可识别的状态来强化视觉构成，这一状态为设计中其他要素的变化与个性提供了基准。从这个意义上说，它不变的统一性容许了构图中其他要素的差异。

第五点秩序原则为"重复"。它可能是最直观的，被广泛应用于不同的形式与情况。建筑中常使用不同类型的重复要素，包括结构（如柱网），空间或诸如窗洞的较小特征。这些重复要素形成了一种功能性的组织结构，并在构成中提供秩序系统[图2.3-19]。在城市层面，重复是城市肌理的固有属性。比如说网格，它是由重复的街道系统分割的、可识别的长方形（或方形）街区模式。其他重复要素可见于城市的基础设施，例如照明设备、人行道、标识、排水渠等。对于强化城市网格的图案与结构，这些要素有着非常重要的作用，虽然它们是相对次一级的角色。

通过重复的要素布置可获得韵律感，它是一种视觉运动暗示。它与重复要素或要素群之间的比例、间距有关。重复由相关联要素组成的群组，就会形成串联的韵律或模式。这一方法在住宅设计中尤为常见：它本质上就是由细小的、可识别要素所组成的重复单元。将不同的韵律按照一个精心协调的顺序合成的构成，通常会拿来与音乐编曲相比较。编曲中常涉及对比鲜明但是切分的韵律，这一点在现代爵士乐中表现突出。当它被转译到建筑中，就会产生一个微妙的、具有视觉刺激的建筑构成，如路易斯·舍特设计的美国驻伊拉克大使馆住宅[图2.3-20]。

disposition within which may exist incidents of *local symmetry*. The program of house for example, usually consists of individual and odd numbers of space requirements (living room, kitchen, dining hall, several bedrooms, etc.). Organizing these elements in a symmetrical plan is a challenge if not impossible. In the design of his home at Monticello, Thomas Jefferson, a classicist, struggled with this dichotomy, only achieving an acceptable resolution with ingenuity and great effort [fig. 2.3-13].

Local symmetry may be effectively employed to resolve conflicts in plan organization as a result of an existing condition such as an irregularly shaped site. A good example is the Hotel Matignon [fig. 2.3-14], a Parisian townhouse of the 18th century. The architect faced a dilemma: how to create a symmetrical formal entry court (cour d'honneur) and a symmetrical façade to a formal garden in the rear that, because of the irregularity of the site, cannot both lie on the same axis. The solution produces two organizations of symmetrically ordered rooms (local symmetry) that overlap and seem to effortlessly connect resulting in a *re-centered axis*.

Hierarchy in a composition creates contrast and order by establishing a difference in ranking or specialness. Compositions lacking in hierarchy may be ordered but most often will be repetitive and monotonous. A checkerboard is an example of a neutral, repetitive arrangement of equal sized squares lacking in hierarchy. One of the most ubiquitous features of modern office buildings, the glass curtain wall, is another example of a well-ordered but ultimately mundane surface lacking hierarchical structure. As an even and featureless continuum, it has in several cases been used as a background *datum* plane with other elements presented in the foreground as unique and special objects (e.g. Salvation Army Building by Le Corbusier, 1932, Paris [fig. 2.3-15]).

There are many ways in which to achieve hierarchy. Size is an obvious characteristic. Shape, position, and detail are all features that can establish difference, and hence, hierarchy. [fig. 2.3-16] An important consideration is that difference is clearly established. The concept of *significant dimension* suggests that a slight increase in size or a minor variation in shape may tend to be perceived as an error or misalignment thereby nullifying the intended contrast and hence, hierarchy. [fig. 2.3-17] Proportional relationships cannot be ambiguous otherwise the composition is unable to express the intended meaning.

Datum is the fourth of the five ordering principles. A datum can exist in several forms including linear, planar and volumetric. In composition, a datum is a structure that provides a common reference for a variety of different and unique elements [fig. 2.3-18]. It establishes order in what may appear to be a random pattern of disparate objects. The degree of randomness may vary from "no discernable pattern" to a repetitive and highly structured arrangement of identical forms. In either case, the datum strengthens the visual composition by providing an unchanging identifiable condition that serves as counterpoint to the variation and individuality of other elements in the design. In this sense its unchanging uniformity allows for the dissimilarity of other elements in the composition.

The fifth ordering principle is that of *repetition*. Perhaps the easiest to visualize, repetition can occur in a wide range of forms and contexts. In architecture, buildings use many types of repetitive elements such as structure (e.g. column grids), spaces or smaller features such as window openings. These repeating elements form a kind of functional organizing structure that provides a system of order in a composition. [fig. 2.3-19] At the urban scale repetition is intrinsic to the patterns of cities. The grid, for example is a pattern of identical rectangular (or square) city blocks separated by a repetitive system of streets. Other instances of repetition are present in urban infrastructure: lighting elements, crosswalks, signage, drainage culverts, etc. These elements play an important although secondary role in reinforcing the pattern and structure of the urban grid.

Rhythm is a type of implied visual movement made possible by the repetition of elements creating a pattern. It involves the proportions and spacing of repeated elements or clusters of elements. The repetition of a group of related elements is referred to as a *concatenated* rhythm or pattern. This device is especially common in housing that is by nature a repetitive unit composed of smaller identical elements. Compositions that combine various rhythms into a carefully coordinated sequence are often compared to musical compositions involving contrasting but syncopated rhythms. This is a feature present in modern jazz that, translated into architecture, creates a subtle and visually stimulating architectural composition (e.g. United States Embassy residence in Baghdad, Iraq by José Luis Sert [fig. 2.3-20]).

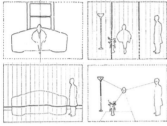

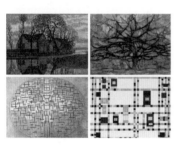

构成

关键概念：平衡，形式与对象内容，抽象，构思，图幅，图形与背景，轮廓线，鲁宾图底关系感知法则，间隙空间，立面构成中的图与底。

鲁道夫·阿恩海姆在《艺术与视觉感知》一书中，将平衡定义为"一种分配状态，其中所有的动作都趋于静止。在一个平衡的构成中，诸如形状、方向与位置的各项因素被同时确定，既无改变的可能，也确保了各组成部分之于整体的必要性。"在绘画中，艺术家通过采用一系列的构图策略来努力实现平衡，包括图幅的划分，重要图形或要素的对位，前景、中景与背景的配置等。在莫奈的绘画作品《拉维科特的塞纳河日落》中，艺术家在画面较中心位置布置中景，由太阳与两座小岛构成一个三角形并框住其中的两艘小船［图2.3-1］。另一个当代的例子是艺术家大卫·霍克尼的作品《亨利·戈尔德撒勒与克里斯托夫·斯科特》。在这幅画中，画家通过布置主要的构图要素来创造一个更佳紧凑、均衡的结构［图2.3-21］。对于这些绘画作品的分析，可以为建筑的形式与构成提供有关形式策略的启示。

抽象对于发展一个批判性的眼光而言是关键的，它使得甄别建筑设计中主要的组织方式与形式成为可能。在1914年到1925年间，皮特·蒙德里安通过一系列作品，实现了从风格化再现方法到纯粹抽象绘画的转变［图2.3-22］。这些艺术作品的内容被消减为纯粹的形式构成。在建筑中，一个设计作品可能源于一个抽象的概念，然后逐渐发展成为一个具有物质性的建造物。从本质上说，平面是建筑物的抽象，它是一个高度规约的图解再现，而非建筑本身。在其概念层面上，建筑的想法可以通过一个简单的形式图解来予以表达，有时候它就是设计构思。

图形之于背景（图底关系）的视觉感知对于空间的理解至关重要。阿恩海姆将闭合轮廓线获得图形属性的现象描述为：显现于无限延伸背景前方的一个闭合界面。在绘画中，外形拮抗（或者说两个闭合图形共享一部分边界）导致了一个变化的、模棱两可的视觉关系。这一现象可见于毕加索的画作《阅读女人的头像》（1953），其中并置的正面与侧面间存在相似片段。这些例子引发了一个疑问：究竟是什么因素决定了图形与背景的阅读？丹麦心理学家爱德格·鲁宾研究了这些现象，并确定了图底关系感知的法则。阿恩海姆对它们进行了阐释与图示，亦可以用它们在艺术作品中的应用加以解释。

与图形感知相关的是间隙空间的状态，即图形之间的背景

composition

Key concepts: *balance, formal versus subject content, abstraction, parti, picture frame, figure and ground, contour line, Rubin's Rules for perception of figure and ground, interstitial space, façade composition related to figure and ground.*

Balance, as defined by Rudolf Arnheim in *Art and Visual Perception*, is "a state of distribution in which all action has come to a standstill. In a balanced composition, all factors such as shape, direction and location are mutually determined in such a way that no change seems possible and the whole assumes the character of 'necessity' in all its parts." In painting, artists strive to achieve balance by employing a range of compositional strategies such as subdivision of the picture frame, alignment of important figures or elements, positioning relative to foreground, middle ground and background, etc. In the Monet painting *Soleil couchant sur la Seine a Lavacourt*, the artist has carefully positioned the sun and the two islands in the middle ground to form a triangle at the approximate center of the painting, also framing two boats floating in this zone [fig. 2.3-1]. A more contemporary painting by the artist David Hockney, *Henry Geldzahler and Christopher Scott*, likewise arranges key elements of the composition to produce a tighter and more balanced structure [fig. 2.3-21]. Analysis of paintings provides insight into formal strategies that are also relevant to architectural form and composition.

Abstraction is essential in developing a critical eye and enables one to distill the primary organization and form in an architectural design. Between the years 1914 and 1925 Piet Mondrian, in a series of paintings, moved form a stylistic representational approach to pure abstract painting [fig. 2.3-22]. Content in the work of art was reduced to its purely formal composition. In architecture, a design may originate as an abstract ideation and develop into a materialized built work. The plan is essentially an abstraction of a building, not an actual work of architecture but a highly conventionalized diagrammatic representation. At its most conceptual, the idea of a building may be characterized as a simplified formal diagram sometimes referred to as the *parti*.

The visual perception of figure versus ground is key to an understanding of space. Arnheim describes the phenomenon of a closed contour line acquiring the attribute of figure, that is, the enclosed surface will appear to be in front of an extended ground plane.

Applied in painting, the idea of contour rivalry, or two enclosed figures sharing a common boundary produces a fluctuating and ambiguous visual relationship. This can be seen for example in a painting by Picasso, Head of a Woman Reading (1953) in which the familiar motif of simultaneous frontal and profile view occurs. In such cases, the question of what factors might determine the reading of a figure or the ground arises. The Danish psychologist Edgar Rubin studied these phenomena and determined certain rules for the perception of figure versus ground. These are explained and illustrated by Arnheim as well as instances of their application in art.

Related to the perception of figure is the condition of *interstitial space*, the ground between figures. It is expected that in painting the space or area surrounding a perceived figure is never left to chance. In classic Greek vase painting the background area (usually black) between figures is carefully manipulated to accentuate and articulate the profiles of the portrayed protagonists and associated objects (mostly rendered in orange). Likewise in architecture, a building that is surrounded on all sides by open space (the building is referred to in this instance as free standing) will be perceived as an object, in which case its contour outline or footprint depicted in plan is a figure. In some cases, the surrounding space may also be configured by adjacent buildings such that it too has figural character. This produces a reciprocal relationship in which the solid (the building mass) and the void (the configured space) may both be perceived as figure, with the dominant reading of figure dependent on the context or other factors (e.g. shape). In urban design, the manipulation of the physical characteristics of the leftover or interstitial space is an important tool for creating a coherent and spatially well-defined city [fig. 2.3-23].

An interesting example of the figure-ground phenomenon in architecture is related to the compositional design of a building façade. Traditional building design was primarily wall construction with window openings as apertures or holes in the surface of the wall. The relative size of windows in relation to the wall is small thereby causing the windows to appear figural with respect to the continuous and unbounded surrounding wall surface. Of course, at the scale of the whole building, the wall has a boundary or edge and is perceived as a field with a figural shape of its own. Nonetheless, we would perceive an area of the wall surface surrounding a window as ground and the window shape as a figure. With the introduction

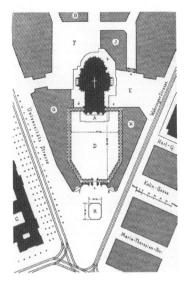

图 2.3-23
卡米诺·西特提出的维也纳沃蒂夫教堂广场调整方案;
rearrangement of the Votive Church Plaza, Vienna proposed by Camillo Sitte;

图 2.3-24
圣·卡塔尔多公墓,
阿尔多·罗西, 1984;
西格拉姆大厦,
密斯·凡·德罗, 1958;
San Cataldo Cemetery, Aldo Rossi, 1984; Seagram Building, Mies van der Rohe, 1958;

部分。通常来说,绘画中应尽量避免使用空间或空白来环绕一个可感知图形。在古希腊花瓶绘画中,通过涂绘图形之间的背景部分(通常为黑色)以强调并清晰表现人物以其附属物件(通常为橙色)的轮廓。在建筑中也是一样:当一栋建筑的周围是开放空间,即独栋存在,该建筑物可被感知为一个物体,其外轮廓或占地范围在平面上表现为图形。很多情况下,与其相临近的,本身也具有图形属性的建筑物会对这些建筑之间的空间定形。这就形成了一个相互关系,实体部分(建筑物体量)与虚空部分(建筑物之间定形的空间)均有可能感知为图形,对于图形的阅读就主要取决于其所处的环境或其他要素,如形状。在城市设计中,操作余留空间或者间隙空间的物理特征,是创造一个连续的、空间上限定清晰的城市的一个重要工具 [图2.3-23]。

在建筑中,有一个有趣的图底关系现象,它与建筑立面构成有关。传统的建筑设计主要采用墙承重结构,窗洞是墙面上的开口。相对于整片墙体来说,窗洞的面积较小,它在相对连续的、无边界的墙体上可感知为图形。当然,就整栋建筑的尺度来说,墙体有其边界范围,可感知为一块以其自身图形为形状的区域。尽管如此,我们还是会将窗洞周围的墙面区域作为背景,窗洞部分作为图形。在框架结构与玻璃幕墙出现之后,图底之间的关系发生了改变。幕墙的窗框网格纤细,反而显得玻璃是连续不间断的。此时,表示窗框的细线网格浮于连续的玻璃表面上,感知网格为图形而玻璃面为背景。[图2.3-24]

在一些建筑中,立面的构成设计呈现出另一种图底关系:墙体与窗洞的比例相对均等。在这种情况下,图底关系的感知就变得不确定或模棱两可。有时候,墙面会表现为背景,窗洞表现为图形,或与之相反,墙面表现为图形,窗洞表现为背景 [图2.3-25]。路易斯·康设计的埃克斯特图书馆立面便是其中一个典型案例 [图2.3-26]。图书馆的外围由砌体承重墙包裹,约一半面积的墙面是由两层高的大块窗户覆盖。另外,随着建筑高度增加,窗洞亦逐渐变大,在顶层屋顶花园位置,墙体消减为柱子经由过梁连接。这里,梁柱部分可阅读为框架,即是图形。在建筑底部情况则发生了反转:地面层的窗框是由拱券限定的洞口,其间的墙面可阅读为连续的背景,与拱券洞口的虚空图形相对。

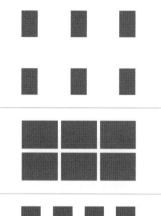

图 2.3-25
窗洞构图与图底关系;
composition of window
openings and figure-ground
relationship;

图 2.3-26
埃克斯特图书馆,
路易斯·康, 1972;
Exeter Library,
Louis I. Kahn, 1972;

of frame construction and the emergence of the glass curtain wall, an inverse of the figure-ground relationship occurs. The curtain wall, with glass seemingly continuous and uninterrupted except by the thin framing grid of the window mullions, presents itself as a continuous glass surface in front of which appears a grid of thin lines representing the mullions. The grid is now perceived as figure and the glass surface as ground. [fig. 2.3-24]

In some buildings, the compositional design of the façade presents a third alternative in which the proportion of wall surface and window is relatively equal. In this case, there is an indeterminate or equivocating perception of figure versus ground. Sometimes the wall surface will appear as ground with the window openings as figure or the reverse, with the wall appearing as figure and the window areas as ground. [fig. 2.3-25] An excellent example of this case is the elevation of the Exeter Library by Louis I. Kahn [fig. 2.3-26]. The exterior envelope of the library is a masonry load-bearing wall. Approximately half of the surface area of the wall is punctuated by large two-story windows that occupy. In addition, as the building rises the window openings increase slightly in width such that at the topmost level of the roof garden, wall surface has been reduced to piers that, together with the lintel spanning between read as frame and hence figure. The reverse occurs at the base of the building. Here the window frames now become the openings in the wall of a ground-level arcade and the wall surface between reads as a continuous ground against the figural voids of the openings of the arcade.

1   Colin Rowe and Robert Slutzky, "Transparency: Literal and Phenomenal," *Perspecta* 8 (1963): 45-54;

2   Rudolf Arnheim, *The Dynamics of Architectural Form* (University of California Press, 1977), 162.

2.4

系统［结构］
system [structure]

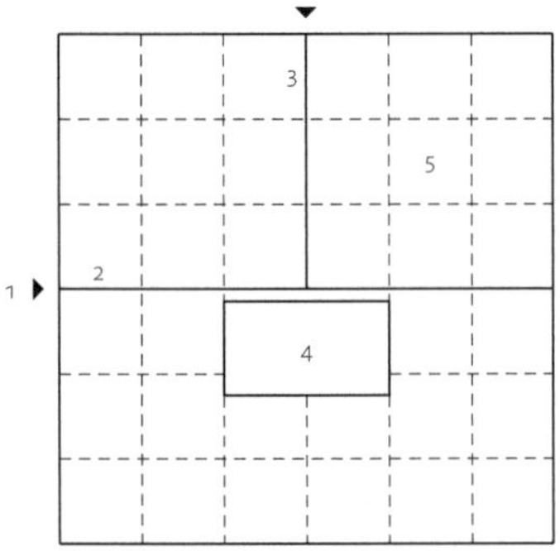

图 2.4-1
罗马古军营网格：
1. 大门; 2. 主干道; 3. 主大街; 4. 广场; 5. 营区;
Roman castrum grid:
1. gate; 2. cardus maximus;
3. documanus maximus;
4. forum; 5. insulae;

图 2.4-2
古罗马城镇提姆加德，位于现在的阿尔及利亚，约公元100年；
Timgad, Roman Town in today's Algeria, ca. 100;

网格

关键概念：模块，契位、移位，变形网格，服务与被服务空间，格纹。

  网格可能是人类所知的一种最古老且无处不在的组织手段。在聚居布局中，使用网格的传统可以追溯到文明的起点。在城市层级上，网格建立了街道与广场网络，为城市创造秩序的同时提供了<u>找路</u>结构［图2.4-1、2.4-2］。在建筑层级上，网格表现为空间与结构的秩序系统，它为布置功能，协调诸如结构与表面饰面的建筑构造要素提供了基本的模式需要。

  网格主要用于场地划分。网格的田字等分是最简单的形式：两条交叉线条，相交位置处的单一节点，四个相互分离的独立空间。网格线将场地划分为多个模数，它们可以表示为三维的空间［图2.4-3］。这些模数或空间单位可以是统一、均等的，如棋盘一般；也可以通过移动网格线的位置来创造变化与层级关系。

  网格是一种基准类型。它可以是中性的，将它想象为无限延伸的场地上有度量的、重复的模数结构，你可以在其中填充、合并网格、余留部分空格，甚至在不损失其基本的秩序组织功能的基础上，进行一定程度的网格变形。在一个网格结构中，事件可能与这个网格模块的尺寸界限相吻合，也可能无视网格的规整而进行自由组织。但不论是哪一种方式，网格所提供的基准允许其整体中存在小部分的差异。

  在建筑中，大部分的网格是由结构建立的，比如说框架结构中的柱网排布。典型的一种情况是按照工程要求，将柱阵依简单、正交的模数网格（比例上可以是方形或长方形）布置。放射状模式是一个特例，它主要服务于平面为圆形或是曲线形的建筑。在一个柱网布局中，通常会将限定空间的垂直板片（分隔墙）与柱子对位布置，这样房间或空间会与网格的模块相一致，我们称这种情况为"<u>契位</u>"［图2.4-4］。然而，由于

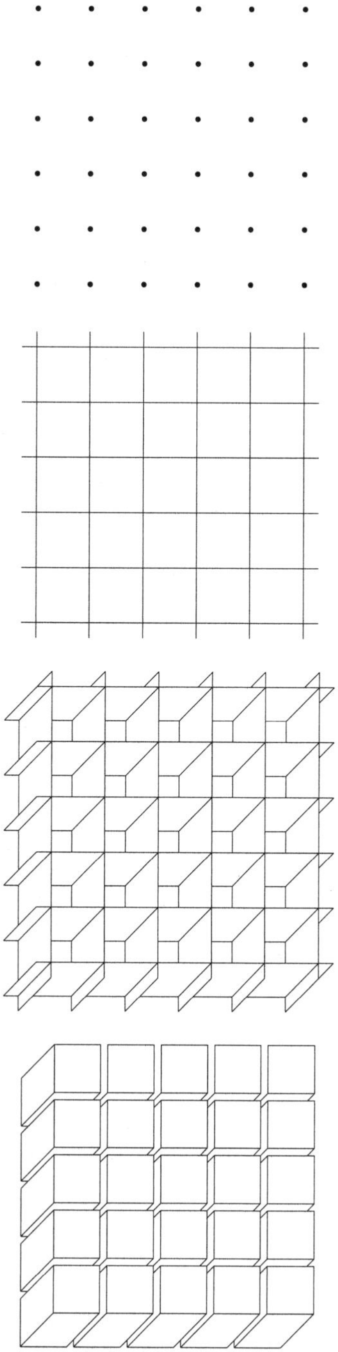

图 2.4-3

网格: 点、线、面、体;
grid: point, line, plane, volume;

grid

Key concepts: *module, close fit, loose fit, free plan, transformed grid, servant and served space, tartan plaid.*

The grid is perhaps the oldest and most ubiquitous organizational device known to mankind. The use of the grid in the layout of human settlements dates back to the dawn of civilization. At the urban scale the grid establishes a network of streets and squares that gives order and *way-finding* structure to the city. [fig. 2.4-1, 2.4-2] At the architectural scale the grid is present as a spatial and constructional ordering system, providing a regular pattern that is required for both functional planning and the coordination of elements of building construction such as structure and systems of surface cladding.

A grid is primarily a subdivision of a field. The four-part subdivision of a square is a grid in its most reduced form: two intersecting lines, a node at their intersection and four separate and distinct areas. Grid lines subdivide a field into modules and these may be interpreted in three dimensions as spatial volumes. [fig. 2.4-3] The modules or spaces may be uniform and equal, as in a checkerboard, or, by shifting the position of the gridlines they may acquire variation and hierarchy.

The grid is a type of datum. It can be neutral, imagined as an extendable field with a measured, repetitive structure of modules that can be filled in, merged together, left as voids, or even distorted to a degree without losing its primary function as an ordering device. Within a gridded structure events may choose to "fit" the dimensional constraints of the grid module or be loosely organized in contrast to the regularity of the grid. Either way, the datum provided by the grid establishes the condition that allows differentiation and individuality to become part of a greater whole.

In architecture, most grids are established by the building structure, for example, the position of the columns in a frame structure. Typically because of engineering requirements columns tend to be arranged in a simple orthogonal grid of modules that are square or rectangular in proportion. An exception is a radial pattern that accommodates buildings that are circular or curving in plan. In a column grid layout the vertical planes (partitions) that define spaces are often aligned with the columns such that the modules of the grid coincide with the rooms or spaces. This is

图 2.4-4

4号住宅，约翰·海杜克，
1954-1963，平面；
House 4, John Hejduk,
1954-1963, plan;

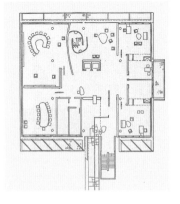

图 2.4-5

棉纺织会总部，
勒·柯布西耶，1954，平面；
Millowners' Association
Building, Le Corbusier,
1954, plan;

图 2.4-6

波森纳斯住宅一，
菲利普·约翰逊，1956，
平面及网格图解；
Boissonas House I,
Philip Johnson, 1956,
plan and grid diagram;

分隔墙本身不承重，在结构上可以选择与柱子分离，这时在就会出现平面布局中不与网格对位的部分或全部分隔墙，这一做法可称为"移位"[图2.4-5]。对应于早期的现代建筑，由它发展出了"自由平面"的开放空间概念。

网格变形是一个进行适应与调整的过程。理想网格所生成的是一组在二维方向上延伸的均一方形模数。而在变形的网格中，模数或空间有可能不是方形的，也由此获得了层级性。网格的变形方式多种多样。通过调整网格线之间的距离，可以改变一个或两个方向上的模数尺寸。这种变形既可以保持其契位模式，也会引发空间规划的多样性。通常，可以根据大、小跨度之间的比例关系来确定网格线的间距，如斯坦因别墅中由柱网限定的大、小区域之比值为1:2。网格变形的其他方式还包括操作网格边界：增减单位网格数、旋转或移动部分网格、合并网格以创造更大的空间。

在菲利普·约翰逊（1906-2005）设计的波森纳斯住宅（1956）[图2.4-6]中，砌筑的结构柱按照5m × 5m的网格布置。柱子的截面尺寸均为500mm × 1000mm，该比例暗示了平面一定的方向性。同时，宽1m的柱子容纳了一系列薄的空间带，可用以布置储藏单元、壁炉及用水服务。为创造一个相对大的无柱起居空间，四个网格单元被合并，并取消中央立柱。同时，为保持住宅整体空间模数比例一致，该部分顶棚高度是其他标准顶棚高度的两倍。

建筑师路易斯·康（1901-1974）于1954年设计了屈灵顿公共浴室[图2.4-7]。这个浴室项目小且建造适度，它在建筑设计上的价值远远超出了其外形。建筑的空间秩序是一个十字形的网格，立柱截面为2.4m见方的空心核心筒，里面布置有诸如盥洗、入口前厅的服务功能。它们支撑着金字塔形的屋顶结构，屋顶下覆盖着跨度6.7m见方的开放空间。康将这些立柱称为"空心的石头"，这种用以布置设备服务的中空结构构件在他之后的设计中被广泛使用。在这一设计特点的基础上，康发展了"服务与被服务空间"的概念。

在这个社区中心项目中，还有一栋未建成的主楼[图2.4-8]，康在该设计中继续发展了其空间秩序概念："以房间创造开间系统"。在这里，决定结构柱位置的网格开始发生变化，以创造可适应不同尺寸房间的大、小模数。这种变形的网格与布料的格纹形式[图2.4-9]相似。

图 2.4-7
屈灵顿公共浴室，
路易斯·康，1954，平面；
Trenton Bath House,
Louis Kahn, 1954, plan;

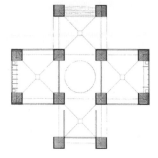

图 2.4-8
屈灵顿犹太人社区中心主楼
（未建成），路易斯·康，
1954，平面图解；
Trenton Jewish Community
Center main building (unbuilt),
Louis Kahn, 1954,
plan diagram;

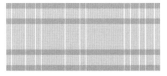

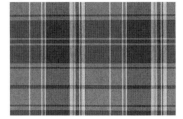

图 2.4-9
格纹布料；
tartan plaid fabric;

referred to as a *close fit* pattern [fig. 2.4-4]. However, since the partition walls are non-load bearing and independent of the columns structurally, it is possible to arrange them in a pattern that is partially or totally non-aligned with the grid. This is referred to as a *loose fit* pattern [fig. 2.4-5]. The *free plan*, an open space concept that was developed in early modern architecture is an outgrowth of the loose fit pattern.

Grid transformation is a process that enables accommodation and adjustment. The ideal grid produces an array of identical square modules extending in two directions. The *transformed grid* is hierarchical in the sense that not all of the modules or spaces are equal. Transformation of the grid occurs in several ways. The spacing of the grid lines can be altered to achieve variation in the dimension of the modules in either one or both directions. This variation creates diversity in space planning while adhering to the close fit pattern. Often the spacing is guided by proportional relationships of big and small bays as in the Villa Stein that has a 1:2 ratio between the small and large zones defined by the column structure grid. Other forms of grid transformation include operations at the edges of the grid field such as the addition or subtraction of grid modules, rotating or shifting portions of the grid, and the joining of modules to create larger spaces.

In the Boissonas House I (1956) [fig. 2.4-6] by Philip Johnson (1906-2005) the masonry structural piers are arranged in a square grid of 5m in each direction. The proportion of the pier, 500mm x 1000mm, implies a directionality to the plan. In addition, the one-meter width of the piers creates thin spatial zones that offer accommodation to storage units, fireplace and wet services. To achieve a large column-free living area four modules are combined into one with the elimination of a pier. The standard ceiling height is doubled thereby keeping the proportions of the space module throughout the house constant.

The architect Louis Kahn (1901-1974) designed the Trenton Bath House in 1954 [fig. 2.4-7]. A small project of modest construction, the Bath House had implications on architectural design far exceeding its appearance. The spatial order of the building is a cruciform grid in which the columns are hollow block cores (2.4m x 2.4m) that contain the services such as the lavatories and entrance vestibules. They support pyramid roof structures over open square spaces (6.7m x 6.7m). Kahn referred to the columns as "hollow

图 2.4-10

原始棚屋，
马克·安东尼·洛吉耶；
primitive hut,
Marc-Antoine Laugier；

结构

关键概念：建筑体系（空间、结构，围合，环境控制与流线）；单元开间，框架系统，格形墙系统，平行墙系统，圆柱－平板系统，自由平面。

在建筑理论中，称人类创造的第一个构筑结构为"原始棚屋"（马克·安东尼·洛吉耶，《建筑论》，1755）[图2.4-10]，它描绘的是一个梁柱木框架：树干为柱，其上由两条木椽（树杈）支撑坡屋顶以形成山墙。这个原始棚屋不单是一个结构，它还是一种理论概念，它提出了建筑中自然、功能与理性的理想原则。它是第一个探讨普适范型之意义与角色的建筑理论，是现代建筑理论所追随的一个先驱。

结构是建筑体系中的一部分，与之并行的还有空间、围合、环境控制与流线。它的角色是基础的：建筑物需要结构方可以站立，在承载荷载的同时还需要安全地抵抗重力与自然力的作用。结构的作用主要有两方面：抵抗重力对其产生的拉力及其支承全部荷载；抵抗倾覆。诸如风、地震、海啸的自然力量可以推倒建筑物。结构需要通过控制刚度、强度与稳定性来抵抗这些水平力的作用。

除了它作为支撑所担负的功能角色外，结构对于建筑形式也有很大的影响。框架系统指的是三维的梁柱网格。框架创造了多个单元开间，它们在尺寸及形状上可能是完全相同的，当然也存在不同情况。网格框架是一个开放、重复的系统，其整齐性创造了秩序，并作为基准存在。在网格框架结构中，可以通过塑造空间来建立空间系统。空间由框架的结构模数限定，可以通过引入垂直与水平的板片来进一步组织空间。框架系统中，这些板片主要是空间限定要素，不具备主要的承重功能。

墙体系统是另一种结构类型。墙体是垂直的板片，其水平向的尺寸通常大于垂直向的尺寸。它们为抵抗荷载而存在，所以在建筑中是主要的结构支撑要素。在历史上有两种正交的墙体系统：格形墙与平行墙。在格形墙系统（有时也称为十字墙结构）中，墙体在正交的两个方向上相互对齐，彼此相交以形成角落。网格单元是由四面墙与四个角落塑造的空间模块。在单向板、双向板或井字梁系统中，横跨于墙上的顶棚与屋顶板同样也是结构元素。

格形墙系统创造了一个强烈限定的、明确的网格。如同网格框架结构一样，创造格形墙网格变化的方法多样：可以改变墙体之间的距离，移除全部或局部的墙体以将两个或多个模块合并，或是改变墙体的正交对位关系（三角形的网格系统就是这种类型的变体）。在大多数的墙体建筑中，都或多或少地采

stones" foreshadowing his future use of structural members with voids that would contain mechanical services. From this design feature Kahn developed his concept of "servant" and "served" spaces.

In the unrealized main building that was part of the overall scheme of the community center [fig. 2.4-8] Kahn further developed the concept of the spatial order as a "bay system of room making". The grid determining the structural column positions now was varied creating large and small modules for rooms of different sizes. This transformation of the grid was identified by its similarity to fabric known as *tartan plaid* [fig. 2.4-9].

structure

Key concepts: *architectural systems: space, structure, enclosure, environmental controls, and circulation; unit bay, frame system, cellular wall system, parallel wall system, round column-flat plate system, free plan.*

The first structure created by man, theorized as the *primitive hut* (Marc-Antoine Laugier. Essai sur L'Architecture, 1755) [fig. 2.4-10], is depicted as a post and lintel frame of wood (tree trunks for columns) with a sloping roof of two wood rafters (tree branches) forming a gable. The primitive hut was more than just a structure, however, it was a theoretical conception that proposed the ideal principles of architecture as natural, functional and rational. It was one of the first architectural theories questioning meaning and the role of universal paradigms, a precursor of modern architectural theory that followed.

Structure is one of the systems of architecture together with space, enclosure, environmental controls and circulation. Its role is essential: buildings need structure to stand up, support loads and safely resist the forces of gravity and nature. Structure does basically two things: it resists the pull of gravity on itself and everything it supports, and it resists falling over. Wind, earthquakes, tsunamis are nature's forces that can topple a building. Structures resist these lateral forces through stiffness, strength and stability.

Besides it's purely functional role as support, structure has a major impact on architectural form. A frame system is a three-dimensional grid of columns and beams. The frame creates *unit bays* that may be identical in size and shape or varying. The grid frame is an open repetitive system that imposes order by its regularity. It acts as a datum. The spaces that are formed within the grid frame structure constitute the space system. Spaces are defined by structural modules of the frame and can be further articulated by the introduction of vertical and horizontal planes. These planes in a frame system are space defining and have no primary load carrying function.

A wall system is a second structural type. Walls are vertical planes with the horizontal dimension normally greater than the vertical. They are designed to resist loads and therefore act as structural support for a building. Historically there are two variations of orthogonal wall systems: *cellular wall* and *parallel wall*. In the cellular wall system (also sometimes referred to as cross wall), walls are

用了变形网格形式。

平行墙系统是网格墙的一种变形。承重墙体按照一个方向相互平行、对位布置。平行墙系统中的网格不如格形墙系统明显。与墙体垂直的网格线可以通过<u>减法</u>——挖去墙体（创造开口或间断）或是<u>加法</u>——增加与墙体垂直的非结构性墙（分隔隔断）来创造。将这些要素对位布置，创造出一个个<u>横穿</u>的空间带，与墙体间重复且平行的空间带形成强烈对比。[图2.4-11]通过改变墙间距、布置分隔墙、开挖洞口，可轻松地操作平行墙系统的网格。由于所有的承重墙均平行布置，横跨其间的顶棚或屋顶板是呈单向板的结构，或也可采用垂直于对边墙体的过梁或板片要素。

表 1

结构类型及其特点；
structural systems and their charateristics；

结构类型 system type	空间特点 spatial character	空间限定 / 结构 space definition / structure
格形墙 **cellular wall**	房间或网格单元 room or cell **传统平面** **traditional plan**	垂直板片作为承重结构 vertical planes as load bearing structure 两个方向上的十字墙 cross-walls in two directions
平行墙 **parallel wall**	连续的方向性空间 continuous directional space **横跨的空间带** **cross-grain spatial zoning**	垂直板片作为承重结构 vertical planes as load bearing structure 间断的平行墙 parallel walls with breaks
框架 **frame**	三维网格框架模数 3D grid frame module **模数的空间平面** **modular space plan**	梁柱结构框架 column and beam structural frame 填充的非结构分隔墙 infill non-structural partitions
板片 **plate**	各方向的连续性 continuous in all directions **自由平面** **free plan**	水平板片（顶棚 / 楼面） horizontal planes (ceiling/floor) 圆柱与无梁板 round columns and beamless slab

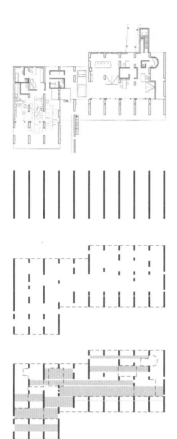

图 2.4-11
萨拉巴依别墅，
勒·柯布西耶，1955，
平面及分析图解（理想网格；
网格变形；横穿的空间带）；
Villa de Madame Manorama
Sarabhai, Le Corbusier, 1955,
plan and analytical diagrams
(idealized grid, transformed
grid, cross-grain spatial
zoning);

aligned perpendicularly in two directions and intersect each other forming corners. Cells are the spatial modules formed by the four walls and four corners. The ceiling or roof plane is also a structural element spanning between walls either as a one-way system of beam elements or a two-way plate or waffle beam system.

Cellular wall systems create a strongly defined and explicit grid. Like the grid frame structure, the cellular wall grid may be transformed by varying the spacing between walls, leaving out walls or portions of walls to combine two or more modules into one, or altering the orthogonal alignment of walls (a triangular grid system is one variant). Most wall buildings have some form of a transformed grid.

The parallel wall system is a variant of the cellular wall. Load bearing walls are aligned in one direction parallel to each other. The grid in a parallel wall system is less explicit than in the cellular wall system. Grid lines perpendicular to the walls are determined either by *subtraction* of the walls to form openings or breaks and by the *addition* of non-structural walls (partitions) perpendicular to the walls. The alignment of these elements creates a *cross-grain* spatial zoning that acts as a powerful contrast to the repetitive parallel spatial zones defined between the walls. [fig. 2.4-11] The grid of the parallel wall system is easily manipulated by the variation in the distance between the walls, the positioning of the partitions, and the location of the breaks in the walls. Because all load-bearing walls are parallel, the ceiling or roof spanning structure is one-way; beam or slab elements span perpendicularly between opposite walls.

The frame system as described earlier, is composed of vertical columns and horizontal beams, connected at nodal points to form a three-dimensional grid frame structure. The column and beam elements in a frame are typically square or rectangular in cross section, making it possible to align the surfaces of the elements with each other. This simplifies the connection of these elements and helps to smoothly articulate their intersection at the node or joint. In addition it improves the alignment of the infill panels, the partitions and ceiling or roof structures that fit either between the columns and beams or against one of their surfaces.

In a column-flat plate system the ceiling or roof-spanning element is a flat, beamless slab, usually of reinforced concrete. Beamless slabs were invented in the early twentieth century (Robert Maillart, Swiss engineer, 1909 patent for a beamless slab) and certain

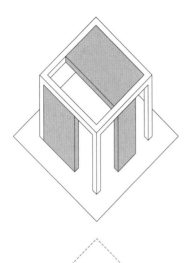

图 2.4-12
框架系统：模数空间；
平板系统：自由平面；
frame: modular space;
flat plate: free plan;

前文所述的框架系统由垂直的柱与水平的梁构成，它们在节点位置相连而形成一个三维的网格框架结构。框架中的梁柱要素在截面上主要是方形或长方形，这使得要素间面与面的对位成为可能。要素之间的连接也因此简化，节点或连接位置亦可平滑的交接。另外，它有助于将板片、分隔墙以及顶棚、屋顶板结构填充于柱或梁之间，或是与它们的某一表面平齐。

在板柱系统中，顶棚或屋顶要素是扁平的无梁板片，材料通常为钢筋混凝土。无梁板片于20世纪初被发明（瑞士工程师罗伯特·马亚尔于1909年取得无梁板的专利），某些现代建筑师迷恋于两片平行、连续的水平板片间的空间潜力。无梁钢筋混凝土板是刚性的横跨结构，需要在其跨度间隔的位置提供垂直支撑，跨度间隔主要取决于板片的厚度。支撑点可以位于板片的任意位置。这一特点就允许设计师在布置垂直支撑要素（柱或墙）时，具有更大的灵活性。平板系统的另一特点是其不具备特定的方向，而表现出中性的状态；空间与运动不受墙体、框架系统中的平行、正交网格的限制。特别是建筑师勒·柯布西耶，他利用无梁板系统的这些特点创造了一种新的空间设计。引入圆形截面的柱来强化空间的多方向性；将分隔墙与梁柱脱离，将其重新自由地排布于空间之中，这与填充墙必须对位于结构要素的布置是完全不同的。它指向了一个开放的、自由的空间布置，即自由平面。[图2.4-12]

modern architects were captivated by the spatial possibilities of two uninterrupted parallel horizontal planes. Beamless reinforced concrete slabs are stiff spanning structures requiring vertical support at some span interval determined primarily by the slab's depth. The point of support can occur at any position of the slab. This characteristic provides the designer with much freedom in positioning the vertical support elements (columns or walls). A second feature of the flat plate system is its non-directional neutrality; space and movement are not constrained by the parallel or orthogonal grids of the wall and frame systems. The architect Le Corbusier in particular exploited these characteristics of the beam-less slab system inventing a new type of space design. Introducing columns of circular cross section that reinforced the multi-direction-ality of space, he also detached the partition panels from their previous infill alignment with the columns and beams and freely re-positioned them in space. This led to an open and flexible space planning called the *free plan*. [fig. 2.4-12]

2.5

体验 [感知]
experience [perception]

氛围

关键概念：光线，漫射光与直射光，光空间，流线路径，找路，入口，门槛，纵贯，空间序列，建筑漫步。

  "光线同时创造了场所的氛围、感觉，以及结构的表现力。"——勒·柯布西耶

  至此为止，我们已经讨论了影响建筑三维存在的形状、配置与形式的诸多形式关系。但是，建筑作为一个人类宜居的人造物，我们对于它的体验主要依赖于移动、触碰、声音与视觉。在这一过程中，我们的感官所接收到的不止是空间的形式与比例，还包括由光线、颜色、材质、温度与声音所细致描绘的空间氛围品质。当这一系列的形式与知觉品质综合在一起时，空间便获得了其独特的性格，我们可以称这种性格或呈现为"氛围"。

  光线是建筑中最为重要的一种氛围品质。路易斯·康曾说过："没有自然的光线，房间便不再是房间。"在设计金博尔美术馆（1972）时，他研究了漫射光品质对于展厅室内空间的影响。在他的设计中，拱形屋顶的顶部位置开有一条连续的天窗，由此引入建筑的直射光通过一个镜面的反射器，改变其方向并照向顶棚。顶棚的拱形形状可以将光线均匀地分散到展厅的垂直墙面上，以利于观众欣赏艺术作品 [图2.5-1]。在一些情况下，建筑师需要在室内空间中引入直射光，以强调某一表面或是创造出夸张的效果。在柯布西耶设计的朗香教堂中，在一个有微光的礼拜堂里，由屋顶天窗引入的一束日光照射在墙面与祭坛上，创造出一种超然的精神气氛 [图2.5-2]。在一个微亮的房间中，夸张的光线可以改变人们对空间的感知。在空间中，可以通过均匀的布置人造光来创造一个整体、统一的光环境（比如超级市场），或是通过聚光来创造一个光空间，即房间中某一区域被光的体积所包裹 [图2.5-3]。这种效果常见于舞台灯光设计中：观众可以清晰洞察舞台上的剧场表演动作，此

图 2.5-2
朗香教堂礼拜堂,
勒·柯布西耶, 1954;
chapel in the Church of
Ronchamp, Le Corbusier, 1954;

图 2.5-3
光的森林, 藤本壮介, 2016;
Forest of Light,
Sou Fujimoto, 2016;

ambience

Key concepts: *light, diffuse versus direct lighting, light space, circulation path, way finding, entrance, threshold, enfilade, spatial sequence, promenade architecturale.*

"Light creates ambience and feel of a place, as well as the expression of a structure." - Le Corbusier

Up to this point we have considered most of the formal relationships that influence the shape, configuration and form of the three-dimensional reality of architecture. But architecture is a *habitable* artifact of man and is experienced through movement, touch, sound, as well as vision. Our senses are receptive not only to the form and proportions of space but also to its ambient qualities nuanced by light, color, texture, temperature, and sound. Space acquires character through the sum of all the formal and sensory qualities it possesses. This character or presence is *ambience*.

Light is the most important ambient quality of architecture. Louis Kahn once said, "a room is not a room without natural light." In his design for the Kimball Museum (1972), he studied the quality of *diffuse* lighting on the interior space of the galleries. A continuous skylight opening at the apex of the roof vault introduces *direct* daylight that is then redirected onto the ceiling by a mirrored reflector. The shape of the vaulted ceiling disperses the light evenly on the vertical surfaces of the gallery for improved viewing of the art works [fig. 2.5-1]. In some situations direct sunlight may be introduced into a space for accentuation of an important surface or for dramatic effect. A dimly lit chapel in the church of Ronchamp by Le Corbusier concentrates daylight from a roof skylight onto the surfaces of the wall and altar creating a transcendent spiritual atmosphere [fig. 2.5-2]. Lighting can also be concentrated within a dimly lit room dramatically changing the perception of the space. Artificial lighting can be evenly distributed in a space to create an overall consistent uniform lighting (such as in a supermarket) or it can be concentrated to create a *light space*, an area within a room that appears to be "contained" by the illuminated volume [fig. 2.5-3]. This effect is common in stage lighting where the action of a theatrical play on stage must be both visible to the audience yet visually restricted.

The architect controls natural lighting through the position, configuration and character of openings in the external envelope, roof, floors and other planar elements. Through various means,

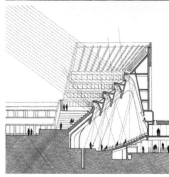

图 2.5-4
阿尔托大学奥塔涅米校区礼堂，阿尔瓦·阿尔托，1966；main auditorium, Aalto University, Otaniemi, Alvar Aalto, 1966；

外的部分视域则受到限制。

通过改变建筑外部围合、屋顶、楼板以及其他片状要素上洞口的位置、配置与特点，建筑师可以对自然光进行控制。通过不同的手段，将自然光引入建筑室内本是黑暗的位置。在平面进深较大的建筑中，庭院、采光井、天井，以及屋顶天窗是获取自然采光的一些常用形式策略。建筑师需要处理建筑的围合界面以控制采光，垂直开口或玻璃覆面可能会造成进入建筑的光线过量或不足。这时候，操控窗洞大小、位置，以及墙体截面上个要素（如挑檐、嵌入式玻璃、百叶或遮阳板等）是获得最佳采光、满足室内光需求的必要手段。［图2.5-4］

英文中的"氛围"一词（ambience）最初来自于法语单词"ambiant"，其原意为"四处移动"。当它由名词转化为动词（ambulate）时，讨论便是空间中的移动。流线对于建筑体验而言是最基本的。人们在空间中的移动与时间有关，它所表示的是一个动态、连续的感受变化。在建筑设计中，对于流线的设计也有很多实际考虑。"找路"一词所描述的是：在空间环境中，跟随或发现路径的现象。建筑中的流线路径或系统应当高效且易于辨认。尤其是紧急疏散通道，应当明确标示。人流系统的形式及其细节特点也应当满足建筑规范中的要求。除了建筑实践上的限制外，人流系统也是建筑形式秩序系统中的一个重要考虑因素。

建筑中的旅程开始于"入口"，每栋建筑都有这么一个区别室内与室外的边界。一栋建筑的入口是一道门槛，作为一个空间转换的位置，它通向的是另一个世界。之所以设置门槛，有一部分原因来自于安保的考虑，即对通道进行控制。此外，门槛还是其守卫领域的特点、价值与身份的象征。这一语义上的角色可以通过具体的设计来传达，包括门洞的大小与形状，门框的形式，视觉通透与否，以及诸如门把手或铰链金属构件的一些细节要素。［图2.5-5、2.5-6］

除了入口之外，建筑的内部空间也应当指引其中的人流移动。在过去，如果人们想从一个房间到另一个房间去，就需要穿过这一空间，并经由门洞进入或离开。考虑到通行的便利性，门洞通常会相互对位布置，也就是法语中的"纵贯"。通常纵贯各门洞的走道是线性、直线形布置。为了避免人流行进至一些不必要去到的房间，建筑师会布置一条单独的流线路径、走道。在一些大型建筑中，往往需要布置很多的走道，它们有时会彼此平行，亦有时相互交叉。走道的布置模式类似于城市的路网系统，在一些情况下可能还具有层级关系。城市中的街道可能指向一个广场或较大的开放空间，建筑中的走道也会通往一些诸如中庭、大厅的开阔室内空间，两者是相类似

图 2.5-5
巴黎圣母主教座堂;
Cathédrale Notre-Dame de
Paris;

图 2.5-6,
母亲之家,
罗伯特·文丘里, 1959;
Vanna Venturi House,
Robert Venturi, 1959;

natural light may be introduced into otherwise un-lite dark areas of the building interior. Courtyards, light wells, atria, and roof lighting are some of the formal devices used to provide natural lighting in a deep-plan building. Light is also controlled at the building envelope: vertical open or glazed surfaces can allow too much or not enough light into the building. The size and position of windows together with the configuration of the wall section (roof overhang, recessed glazing, louvers or brise soleil, etc.) can be manipulated to obtain the best lighting for the requirements of the interior space. [fig. 2.5-4]

Ambience is derived from the French word *ambiant* meaning "going around". From it we get the verb to *ambulate* meaning to move about. *Circulation* is fundamental to our experience of architecture. We move through space in time, a dynamic and constantly changing impression. Circulation also has a practical, utilitarian purpose. *Way finding* is a term that describes the process of following or finding a path through a spatial environment. In a building, the circulation route or system should be efficient and legible. Emergency egress must be clear. In addition, the form and all detail characteristics of movement systems must meet building code requirements for the physically impaired. But aside from the practical constraints, movement systems are also a critical element in the formal ordering systems of architecture.

The journey through a building begins at the entrance. Every building has a boundary that demarcates interior from exterior. The entrance to a building is a *threshold*, a place of transition and a portal into another world. Aside from controlling passage (the function of security), the threshold signifies the character, values, and identity of the realm beyond. This semantic role is expressed by its design, from the size and shape of the door, the framing elements, its transparency or lack of, to details such as the door handle or the hardware of the hinges. [fig. 2.5-5, 2.5-6]

Beyond the entrance, one must be guided through the spaces of the building. In the past, one might go from room to room, passing through each space and through the doorways leading into and out of each room. For convenience the doorways were often in alignment, a condition the French call *enfilade*. To avoid passing through every room including those not required or necessary, a separate circulation path, the corridor was invented. Corridors like enfilade doorways, are normally linear and straight. In large, complex buildings there may be many corridors; sometimes parallel, some-

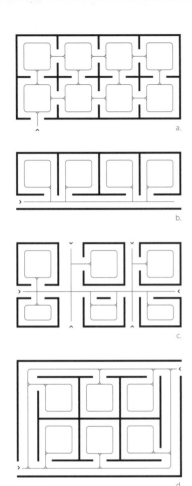

图 2.5-7

流线系统：a. 串联，b. 单
廊，c. 街道网格，d. 外围；
circulation system: a. enfilade,
b. single corridor, c. street grid,
d. periphral access;

的。基本的走道布局形制包括：正交网格、放射状网络、分支
网络以及外围联通 [图2.5-7]。

综合组织流线与空间是设计中的一项基本挑战。[图2.5-8]
对于不同系统类型的选择与组合往往取决于以下因素：建筑的
形制、空间的类型与功能排布、不同的流线表现要求（如自然
采光）等。一些叙事想法也可能会影响到流线的设计。这时
候，人在空间中的移动可能与庄严性、仪式感的表达有关，其
序列的特点与氛围便至关重要。

流线同时也是建筑体验的一部分，它是按照一条线路展
开。这一想法的本质就是建筑师勒·柯布西耶所提出的"建筑
漫步"概念 [图2.5-9]，该概念在其建筑设计与城市规划中均有
广泛应用。"体验建筑，是一个人漫步其中并穿行而过……所
以，决定一个建筑作品是死寂或生动，主要在于身处其中的人
们是否能感受到漫步的法则。"[1]

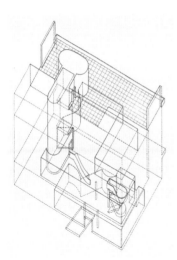

图 2.5-8
斯坦因别墅流线/空间序列图
解分析；
analytical study of circulation/
space sequence, Villa Stein;

图 2.5-9
叙事图像笔记，勒·柯布西耶
致迈耶夫人的信；
narrative visual essay,
letter from Le Corbusier to
Madame Meyer;

times intersecting. The pattern of corridors may resemble an urban street grid and may also be hierarchical. In the same way that a street may lead into a plaza or large open space, corridors sometimes flow in large interior spaces such as an atrium or hall. Basic configurations of corridor layouts include: orthogonal grid, radial network, branching network and peripheral access [fig. 2.5-7].

The integration of circulation and spatial organization is one of the fundamental challenges in design. [fig. 2.5-8] Selection of the appropriate type or combination of systems depends on many parameters including the building configuration, the type and functional arrangement of the spaces, various performance requirements (e.g. natural lighting) of the circulation, etc. Narrative ideas might also influence the design of circulation. Movement might be configured as an expression of solemnity or ceremony, the ambience of the character of the sequence being paramount.

Circulation might also be conceived as a device for experiencing architecture as it unfolds along a path. This is essentially the concept of the *promenade architecturale* [fig. 2.5-9], introduced by the architect Le Corbusier in both his building designs and urban planning. "Architecture is experienced as one roams about in it and walks through it... So true is this that architectural works can be divided into dead and living ones depending on whether the law of 'roaming through' has not been observed or whether on the contrary it has been brilliantly obeyed." [1]

1   By Le Corbusier (1942), cited from Flora Samuel, *Le Corbusier and The Architectural Promenade* (Birkhauser, 2010), back cover.

参考书目 | key references

下列文献可以为课程介绍的主题与内容提供额外的阅读资料。
The following references provide additional reading on the topics and content of the course.

技法 [绘图] technique [drawing]

1. Francis D. K. Ching, *Architectural Graphics*, 6th. ed (Hoboken, New Jersey: John Wiley & Sons, Inc, 2015). [美] 程大锦, 张楠, 张威. 建筑绘图(第6版) [M]. 天津: 天津大学出版社, 2019.

2. Jonathan Block Friedman and David Diamond, *Creation in Space: A Course in the Fundamentals of Architecture. Vol. 1*, 2nd. ed (Dubuque, Iowa: Kendall/Hunt, 2000).

3. Paul Laseau, *Graphic Thinking for Architects & Designers*, 3rd. ed (New York: Wiley, 2001). [美] 拉索, 邱贤丰. 图解思考: 建筑表现技法(第3版) [M]. 北京: 中国建筑工业出版社, 2002.

4. Norman A. Crowe and Steven W. Hurtt, "Visual Notes and the Acquisition of Architectural Knowledge," *Journal of Architectural Education* 39, no. 3 (April 1986): 6–16, https://doi.org/10.1080/10464883.1986.10758395.

5. Michael Graves, "The Necessity for Drawing," in *Michael Graves* (New York, NY: Princeton Archit.Press, 2005), 235–45.

形式 [要素] form [elements]

6. Jacqueline Gargus, *Ideas of Order: A Formal Approach to Architecture* (Dubuque, Iowa: Kendall/Hunt Pub. Co, 1994).

7. Roger Sherwood, *Principles and Elements of Architecture* (Los Angeles: School of Architecture, University of Southern California, 1985).

8. Francis D. K. Ching, *Architecture: Form, Space, & Order*, 4th. ed (Hoboken, New Jersey: Wiley, 2014). [美] 程大锦, 刘丛红. 建筑: 形式、空间和秩序(第4版) [M]. 天津: 天津大学出版社, 2018.

构成 [空间] composition [space]

9. Rudolf Arnheim, *Art and Visual Perception: A Psychology of the Creative Eye*, expanded and rev. ed (Berkeley: University of California Press, 2009).

10. Bruno Zevi and Joseph A. Barry, *Architecture as Space: How to Look at Architecture*, rev. ed (New York: Da Capo Press, 1993).

11. Colin Rowe, Robert Slutzky, and Bernhard Hoesli, *Transparency* (Basel ; Boston: Birkhäuser Verlag, 1997). [美] 柯林·罗, [美] 罗伯特·斯拉茨基, 王又佳, 金秋野. 透明性 [M]. 北京: 中国建筑工业出版社, 2008.

12. Colin Rowe, *The Mathematics of the Ideal Villa, and Other Essays* (Cambridge, Mass: MIT Press, 1976).

13. Steven K. Peterson. "Space and Anti-space". In Jeffrey Horowitz ed., *Harvard Architectural Review 1: Beyond the Modern Movement* (Cambridge: MIT Press, 1980): 89-113.

系统 [结构] system [structure]

14. Edmund N. Bacon, *Design of Cities*, rev. ed (New York: Penguin Books, 1976).

15. Roger H. Clark and Michael Pause, *Precedents in Architecture: Analytic Diagrams, Formative Ideas, and Partis*, 4th. ed (Hoboken, N.J: John Wiley & Sons, 2012).

16. Jeffrey Balmer and Michael T. Swisher, *Diagramming the Big Idea: Methods for Architectural Composition*, 2nd ed. (New York: Routledge, 2019).

体验 [感知] experience [perception]

17. Pierre von Meiss, *Elements of Architecture: From Form to Place* (London ; New York, NY: Van Nostrand Reinhold, 1990).

18. Flora Samuel, *Le Corbusier and the Architectural Promenade* (Basel: Birkhäuser, 2010).

2.1-1   Christian Kerez, "Hermitage, Oberrealta," 2G International Architecture Review, no. 14 (2000): 124;

2.1-3   photo & drawings © Richard Meier & Partners Architects, www.richardmeier. com/?projects=shamberg-house-2;

2.1-5   Anthony Vidler, James Frazer Stirling: notes from the archive (Yale University Press, 2010), 131-132;

2.1-6   Auguste Choisy, Histoire de L'Architecture, Tome II (Paris: Gauthier-villars,1899), 49;

2.1-7   Vidler, Stirling, 137;

2.1-8   drawing © Steven Holl Architects, www.stevenholl. com/projects/stretto-house;

2.1-9   photo © Iggi Falcon, www.flickr.com/photos/ bautisterias/9212182042/; drawing: www. theartroomonline.net/2014/06/cityscapes-ambrogio-lorenzetti.html;

2.2-1   photo © Dennis Jarvis, www.flickr.com/photos/ archer10/2216719429

2.2-2   photo © I-Ta Tsai, www.flickr.com/photos/tsaiid/ 12445312405/; drawing: quod.lib.umich.edu/h/ hiaaic/x-BF70AB/BF70AB

2.2-3   omeka.wellesley.edu/piranesi-rome/exhibits/show/ romanobelisks/flaminian

2.2-4   www.ethiopianadventuretours.com/ethiopia-travel

2.2-5   selfietecture.com/classic/

2.2-6   Museum of Modern Art (New York, N.Y.), ed., Five Architects: Eisenman, Graves, Gwathmey, Hejduk, Meier (New York: Oxford University Press, 1975), 46.

2.2-7   Rudolf Arnheim, Art and Visual Perception: A Psychology of the Creative Eye (Berkeley: University of California Press, 1974), 13.

2.2-8   Ibid., 10, 12.

2.2-12   Wittkower, Architectural Principles, 69.

2.2-14   Le Corbusier et al., Œuvre complète . Bd. 1: 1910 - 1929 (Basel: Birkhäuser, 1999), 144.

2.2-15   Ralph Johannes, ed., Entwerfen: Architektenausbildung in Europa von Vitruv Bis Mitte Des 20. Jahrhunderts: Geschichte, Theorie, Praxis (Hamburg: Junius, 2009),140.

2.2-17   Le Corbusier, The Modulor: A Harmonious Measure to the Human Scale, Universally Applicable to Architecture and Mechanics (Basel ; Boston: Birkhäuser, 2000), cover, 81.

2.3-1   paiting © Petit Palais, musée des Beaux-arts de la Ville de Paris, parismuseescollections.paris.fr/fr/ petit-palais/oeuvres/soleil-couchant-sur-la-seine-a-lavacourt-effet-d-hiver.

2.3-4   photo © Peter J. Sieger, www.siegerarchphoto.com/

2.3-5   photo: galleries.apps.chicagotribune.com/ chi-20140420-frank-lloyd-wright-homes-battle-pictures/; drawing: William Allin Storrer, The Frank Lloyd Wright Companion (Chicago: University of Chicago Press, 1993), 15.

2.3-6   Ibid., 101.

2.3-7   Tracy Elizabeth Cooper and Andrea Palladio, Palladio's Venice: Architecture and Society in a Renaissance Republic (New Haven: Yale University Press, 2005), 242, 240.

2.3-8   photo © Kristian Adolfsson, architecture photography.nu/wp-content/uploads/sites/4/2019 /10/Bauhaus-Dessau-Ro%C3%9Flau-Germany-Deutschland-Walter-Gropius-2018-bw-01.jpg

2.3-9   Le Corbusier et al., Œuvre complète . Bd. 4: 1938 - 1946 (Basel: Birkhäuser, 1999), 55.

2.3-10   paiting © Museo Reina Sofia, artsandculture.google. com/asset/grapes-juan-gris/OgFxw4qV3i5VxQ

2.3-11   Le Corbusier et al., Œuvre complète . Bd. 1: 1910 - 1929 (Basel: Birkhäuser, 1999), 142.

2.3-13   I.T. Frary, Thomas Jefferson Architect and Builder (Garrett & Massie, 1931), plate I, XXV.

2.3-14   Michael Dennis, Court & Garden: From the French Hotel to the City of Modern Architecture (Cambridge, Mass: MIT Press, 1986), 102.

2.3-15   Le Corbusier et al., Œuvre complète . Bd. 2: 1929 - 1934 (Basel: Birkhäuser, 1999), 97, 108.

2.3-16   www.archweb.it/dwg/arch_arredi_famosi/ James_Stirling/Centro_Olivetti/Olivetti-Training-Centre-Haslemere.jpg

2.3-19   Stuart Wrede, The Architecture of Erik Gunnar Asplund (Cambridge, Mass: MIT Press, 1980), 134.

2.3-20   Knud Bastlund, José Luis Sert: Architecture, City Planning, Urban Design (Birkhäuser Basel, 1967), 104.

2.3-21   hragvartanian.com/2009/07/10/more-henry-geldzahler/.

2.3-22   John Milner and Piet Mondrian, Mondrian, 1st pbk. ed (London: Phaidon, 1994), 65, 98, 119, 220.

2.3-23   Translation of City Planning according to Artistic Principles (Camillo Sitte, 1889), in George R. Collins et al., Camillo Sitte: The Birth of Modern City Planning (New York: Rizzoli, 1986), 282.

2.3-24   photo of San Cataldo Cemetery © Laurian Ghinitoiu, www.archdaily.com/95400/ad-classics-san-cataldo-cemetery-aldo-rossi; photo of Seagram Building © Maciek Lulko, www.flickr.com/photos/ lulek/45180370294/in/photostream/

2.3-25   diagram by author, reference: Rudolf Arnheim, Art and Visual Perception: A psychology of the Creative Eye (Berkeley: University of California Press, 1954), 241.

2.3-26   Joseph Rykwert and Louis I. Kahn, Louis Kahn (New York: H.N. Abrams, 2001), 122.

2.4-2   drawing © Frederik Pöll: socks-studio. com/2017/06/21/a-perfect-grid-the-roman-town-of-timgad-the-african-pompeii/

2.4-4   John Hejduk, John Hejduk, 7 Houses, Catalogue - IAUS 12 (New York, N.Y: Institute for Architecture and Urban Studies, 1979), 63.

2.4-5   Le Corbusier et al., Œuvre complète . Bd. 6: 1952 - 1957 (Basel: Birkhäuser, 1999), 149.

2.4-6   Stover Jenkins, Philip Johnson, and David Mohney, The Houses of Philip Johnson (New York: Abbeville Press, 2001), 154-155.

2.4-7   Romaldo Giurgola and Jaimini Mehta, Louis I. Kahn (Boulder, Colo: Westview Press, 1975), 116.

2.4-8   Ibid., 115.

2.4-9   www.kovifabrics.com/search/product_detail/8240

2.4-10   Marc Antoine Laugier, Wolfgang Herrmann, and Anni Herrmann, An Essay on Architecture, Documents and Sources in Architecture, vol. 1 (Los Angeles: Hennessey & Ingalls, 1977), xxiv.

2.4-11   Le Corbusier et al., Œuvre complète . Bd. 6: 1952 - 1957 (Basel: Birkhäuser, 1999), 118.

2.5-1   photo © Peter J. Sieger, siegerarchphoto.com/ kimball-art-museum; drawing: Luca Bellinelli, Louis I. Kahn, and Museo d'arte Mendrisio, eds., Louis I. Kahn, the Construction of the Kimbell Art Museum (Milan : New York: Skira editore): 100;

2.5-2   archhistdaily.files.wordpress.com/2012/05/ronchamp. jpg;

2.5-3   images.adsttc.com/media/images/570e/2504/e58e/ cef2/f400/015a/large_jpg/copyright_laurian_ ghinitoiu_fujimoto_(1_of_31).jpg?1460544765;

2.5-4   photo © Alvar Aalto Museum/Martti Kapanen, www. domusweb.it/en/biographies/alvar-aalto.html; drawing: G. Z. Brown and Mark DeKay, Sun, Wind, and Light: Architectural Design Strategies (Hoboken: Wiley, 2013), ii.

2.5-5   www.curbed.com/2019/8/21/20813675/ vanna-venturi-house-philadelphia-owner

2.5-6   photo © Markus Brunetti, www.yossimilo.com/artists/ markus-brunetti#&gid=1&pid=markus_brunetti-paris_cathedrale_notre_dame-5

2.5-9   drawing © Paris, Foundation Le Corbusier.

3 练习 exercises

3.1 技法［绘图］练习
exercises for technique [drawing]

3.1.1
徒手线条 freehand drawing

任务书 handout

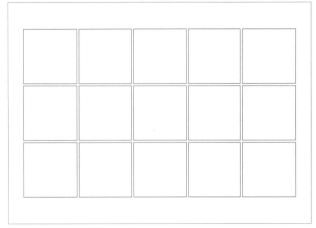

图纸布局, drawing layout;

该练习的目的在于，训练学生准确绘制高品质徒手直线的能力。学生可以通过反复练习，并使用正确技巧来掌握这项基本的绘图技能。

训练要求

线条除了描述走势的直曲外，还包括其他很多特征，如宽度（即线宽），深度（从灰到黑）以及式样（实线、虚线、点线等）。当不同特征组合在一起，就可以创造出多种多样的线条类型。绘制线条时，不论选择哪种媒介，组合使用不同类型的线条将有助于清晰的图纸表达。

练习使用一张标准的A3白色绘图纸，将其水平横向放置。使用很轻的线条在绘图纸中央排布一组横5格、纵3格的方格网，方格70mm x 70mm，彼此间隔5mm。这组方格网距离纸张对边的距离相同。也就是说，绘图纸上与下、左与右的留白相等。因方格网仅作为参考网格使用，绘制时线条应尽可能轻。

线条绘制从左上角的方格开始（对于左利手的同学可以从右上方的网格开始），首先使用深且利索的轮廓线来绘制网格边框。然后按照下述给定的线条类

The intention of this exercise is to develop the ability to draw freehand straight lines of precision and good quality. Basic freehand drawing skills can be acquired with practice and correct technique.

REQUIREMENTS

In addition to being straight, a line can have certain characteristics such as width (commonly referred to a line weight), darkness (gray to black), and pattern (solid, dash, dot, etc.). These features can combine in various ways to create a wide variety of line types. No matter what media is used to create a drawing, the use of different line types help to give the drawing greater definition and clarity.

Begin with a single A3 sheet of good quality white drawing paper. Orient the sheet horizontally. Now using very light guidelines, draw a 3 x 5 grid of squares centered on the sheet. Each square should be 70 mm on a side and be separated from the adjacent squares by 5 mm. Centering the field of squares implies equal borders on opposite sides of the sheet. In other words, the top and bottom margins should be equal as well as the right and left margins.

Begin by outlining in freehand the upper left square (for left-handed individuals, begin with the upper right square.). Use a dark, sharp profile line.

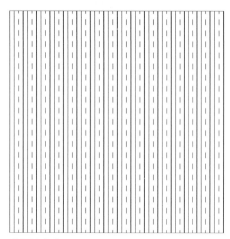 

绘图图案（一）, drawing pattern I;

型组合模式，在方格中填布徒手线条。线条之间应保持1.5mm的间隔。在训练过程中，可使用刻度尺度量以控制线条之间的间距，直到能够摆脱度量工具准确地控制线条间距为止。

　　在第一行方格中，由左及右依次徒手填布水平直线线条。线条绘制需依照下述模式交替进行：
　　－ 第一条：剖切线—实线（粗线）
　　－ 第二条：轮廓线—实线（中等深度，挺括）
　　－ 第三条：轮廓线—虚线（同上）
　　－ 第四条：立面线—实线（浅，但锐利）
　　－ 第五条：立面线—虚线（同上）

　　在第二行的方格中填布同心方格或圆。在左侧第一个方格中绘制同心方格，最外圈使用粗线绘制，然后依次向内圈发展，线条亦随之逐渐变轻变细，每圈方格之间需保持1.5mm间隔。在左侧第二个方格中采用相反模式，即在最外圈使用最轻最细的线条，向内圈发展时线条逐渐加重。在第三个网格中，仅使用轮廓线—实线（中等深度，挺括）绘制同心圆，彼此间隔为1.5mm。右侧的两个方格则镜像重复左侧方格相同的绘图模式。

Now begin to 'fill' the squares with lines based on the pattern indicated below. These lines are to be spaced 1.5mm apart. Measure the spaces with a scale until you are able to determine the spacing correctly without measurement.

Beginning with the left-hand square in row 1, draw lines horizontal to the sheet. In this row you are to alternate the type of line according to the following pattern:

- First line: section cut (Bold)
- Second line: profile (Medium dark, crisp)
- Third line: profile dash (same)
- Fourth line: elevation line (light but sharp)
- Fifth line: elevation line dash (same)

In the middle row of squares, starting at the left, draw concentric square rings spaced 1.5mm apart. In the first square, use the bold line for the outside square and then gradually make the lines lighter and lighter as you draw smaller squares. In the second square, reverse the pattern and use a light line for the outside square, then gradually darker towards the center. For the third square, the center position of the grid, draw circles again spaced 1.5mm apart. Make these lines the same line weight; profile lines dark and crisp. In squares 4 and 5 repeat the pattern of the squares 1 and 2

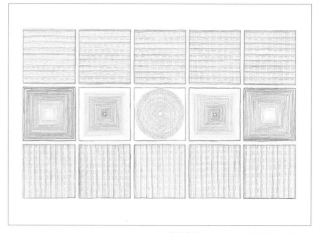

学生作业, student work (LEUNG, Ho Wai Ian);

在最下面一排的方格中，重复应用第一排的绘图模式，只是将水平线条换成竖向的垂直线条。

为获得最佳的绘图效果，绘制线条时应尽可能一笔连续完成。及时削尖铅笔笔头，以保持线宽一致。在练习过程中，学生可尝试使用不同硬度的铅笔，如F，HB，2H，2B等。绘图过程中，应尽量避免使用橡皮擦进行修改。

评价标准
- 准确度量及排布的方格网
- 线条品质：线形的变化与线宽的控制
- 线条间距、平直程度以及准确度

(but mirrored).

In the bottom row of squares, repeat the pattern of the first row except substitute vertical lines for horizontal lines. Use the alternating pattern of line types.

To attain the best results, try to draw each line in a single continuous motion. Sharpen the pencil frequently. Experiment with different leads: F, HB, 2H, 2B, etc. Strive for consistency. Avoid erasing lines.

EVALUATION CRITERIA
- Correct measured construction of grid
- Line quality: variation of line types and consistency of line weight
- Spacing, straightness and precision of lines

变式 variation

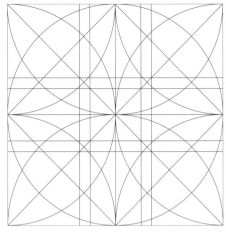

练习使用一张标准的A3白色绘图纸，将其水平横向放置。使用很轻的线条在绘图纸中央排布一组横5格、纵3格的方格网，方格70mm × 70mm，彼此间隔5mm。这组方格网距离纸张对边的距离相同。也就是说，绘图纸上与下、左与右的留白相等。因方格网仅作为参考网格使用，绘制时线条应尽可能轻。

上方所示的绘图图案与正方形的几何性质相关。练习以此图案作为参照，从中选择线条，然后徒手重绘以填补这个15个方格。每个方格中，连续线条（不论直曲）的数量不得少于12。在绘图前，你需要对该图案的逻辑进行研究，它会提示你对于线条的选择。另外，绘图时应组合使用不同的线条类型，如剖切线—实线（粗）、轮廓线—实线/虚线（中等深度，挺括）及立面线—实线/虚线（浅，但锐利）。由此各方格中的图案才能彼此区分，同时，创造出一个独一无二的构图。

绘制线条时应集中注意力，尽可能一笔连续完成。及时削尖铅笔笔头，以保持线宽的一致。绘图可使用4B（剖切线），2B（轮廓线）与H（立面线）3种硬度的铅笔，尽量避免使用橡皮擦修改。

One A3 sheet of white drawing paper. Orient the sheet horizontally. Now using very light construction lines, draw a 3 x 5 grid of squares centered on the sheet. Each square should be 70 mm on a side and be separated from the adjacent squares by 5 mm. Centering the field of squares implies equal borders on opposite sides of the sheet. In other words, the top and bottom margins should be equal as well as the right and left margins.

Now, refer to the diagram above which represents geometric patterns that are generated from the properties of the square. From this pattern select lines to redraw within the fifteen squares. There must be no fewer than 12 continuous freehand lines (straight or curved) per square. Before you start, you should study the logic of the pattern above. This may inform your selection of the lines you will draw in the squares. In addition, use a variety of line types, such as section cut (bold), profile (medium dark, crisp), profile dash (medium dark, crisp), elevation line (light but sharp), elevation line dash (light but sharp), such that the patterns of lines in the squares are further distinguished from each other while, at the same time, creating a unified composition.

Concentrate on drawing lines in one continuous movement. Sharpen the pencil frequently. Strive for consistency. Use 4B (section cut), 2B (profile) and H (elevation). Erasing should be avoided.

小结 reflection

在艺术与建筑的学习中，线条绘制都是一项传统的初步训练内容。在没有直尺的帮助下得心应手的控制线条绘制，其重要性在于提升绘制徒手线条的能力，这是建筑师的一项基本技能。在训练中，使用正确的技法，不仅可以增加学生对于手绘的信心，还可以提升徒手草图的品质。

练习的第一步，要求在A3绘图纸的中心位置，使用绘图工具（如丁字尺，三角板与比例尺）来绘制网格，进行图面布局。这可以训练初学者正确使用工具及准确度量的方法。其次是构建整体的辅助线。练习的过程同绘制建筑图纸的步骤思想一样: 先从全局入手，在掌握了其基本模式后，再深入处理各特定的细节。

除了控制线条（即在两点之间笔不离纸的连续绘制一条直线）外，这个练习对于线条品质也提出了要求。连贯性与线宽是获得线条品质的两个关键因素。连贯性要求线条宽度均匀、挺直。它也指控制线条间距以及虚线线段间的间隔。线宽是用笔力度与铅笔笔芯类型（如2H或4B）选择相协作的结果。线宽有助于区别不同的线条类型。

Line drawing is a traditional beginning exercise in both art and architectural studies. The importance of developing control in drawing lines without straightedge tools is that it improves freehand drawing technique, a fundamental skill for architects. Practice using correct technique creates confidence in the ability to draw by hand and improves the quality of the freehand sketch.

The first step in the exercise is to construct a grid of boxes centered on an A3 sheet of drawing paper. This layout construction is made using drawing tools (T-square, triangle and scale) giving the beginning student practice in both the correct use of drawing instruments and precise measurement. The process of drawing the overall construction guidelines first, instills the approach of beginning an architectural drawing from the general overall pattern and then proceeding to the specific or detail.

In addition to line control, that is, being able to draw a straight line from point A to B without lifting the pencil, the exercise also requires a demonstration of *line quality*. Line quality is achieved with *consistency* and *weight*. Consistency refers to an even line width and straightness. It also refers to consistent spacing of the line gaps and segments of a dashed line. Weight is a combination of pressure and drawing lead type (e.g. 2H or 4B). The weight of a line helps to differentiate different *line types*.

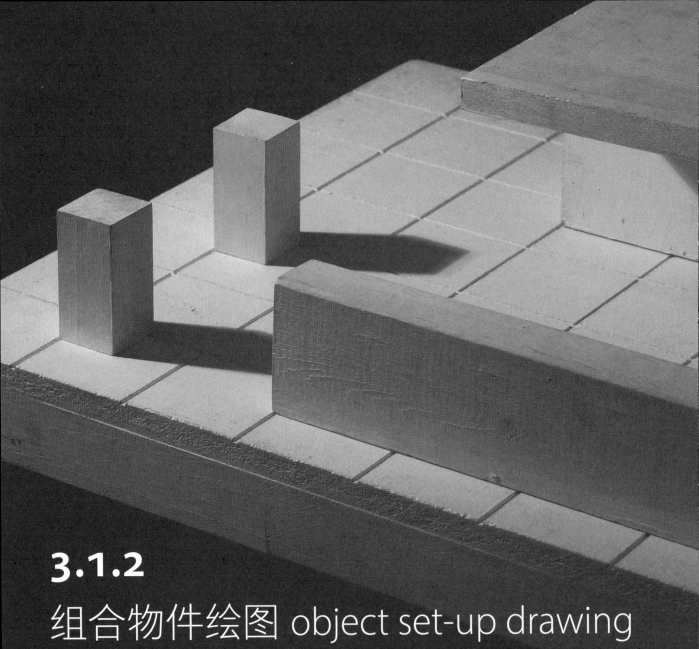

3.1.2
组合物件绘图 object set-up drawing

任务书 handout

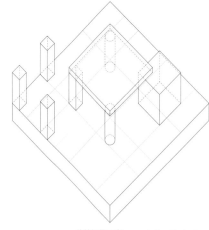

物件摆放示意, example for object set-up;

练习要求按照投形与轴测的图示方法，记录一组给定的形式组织模型。另外，该练习是对学生绘制徒手线条能力的检测，并要求重视模数、比例以及线条品质。

训练要求

在给定的A3绘图纸模版及描图纸上，徒手绘制物件模型的平面图、立面图及剖面图。练习中，学生需要选择合适的剖面剖切位置，以尽可能多地表达该物件组合的信息。平面图与剖面图都是通过剖切物件获得，差别在于两者剖切平面的方向。在图纸上（如绘图纸模板所示），将立面图摆放在平面图之下，可通过其对位及朝向关系来反映它们之间的关联。而剖面图则可以由平面图上的剖切符号获得暗示。

训练中使用的物件模型采用了统一的模数。在绘图中，通过比较各部件要素与底板网格之间的模数关系（1:1，1:1.5，1:2等）以准确掌握各物件的相对尺寸。所以在绘图纸上不强调具体的比例尺。

然后，取一张A3大小的描图纸用以绘制物件模型的轴测图。使用先前绘制的平面图为衬底，将平面图

The intention of this exercise is to graphically describe an arrangement of form using the conventions of planimetric and axonometric drawing. Furthermore, this assignment tests freehand drawing skills and requires careful attention to module, proportion and line quality.

REQUIREMENTS

Plan, elevation and section drawings of the study model. Freehand pencil on A3 formatted drawing sheet and tracing paper. Select the cutting planes that will reveal the most information about the object. Remember that a plan is a section cut parallel with the ground. One elevation should be drawn beneath the plan (as indicated on the format sheet) and relate to it by position and orientation. And cut signals can be used to indicate the relationship between section and plan.

Remember there is no scale given for the construction of these drawings. However, the elements are modular to each other and the grid of the base. The module (1x1, 1x2, 1 ½ x 1 ½, etc.) can be used to draw the elements in correct proportion.

Axonometric drawing of the study model. With a sheet of A3 tracing paper draw/construct the axonometric using the plan drawn above as an underlay. In other words, project the plan vertically in order to create the

绘图纸模版, formatted drawing sheet;

中的部件要素垂直向上投形，来表达其三维信息。利用剖面图及立面图来获取这些部件要素的高度尺寸。与平、立、剖面图一样，轴测图的表达亦不强调所使用的比例尺。轴测图需排布在A3描图纸的中央位置（图纸方向不限）。

评价标准

- 正确以及全面的信息表达
- 模数、比例以及形状的准确
- 线条品质与线宽的把握
- 平面图、剖面图与轴测图之间的关系

third dimension. Use the section and elevation to obtain the lengths of the vertical lines. Again, there is no scale given for the construction of this drawing. It should be centered on an A3 sheet of tracing paper (vertical or portrait orientation).

EVALUATION CRITERIA

- Correct and complete information
- Module, proportion and shape
- Line quality and line weight
- Relationship of plan/section/axon

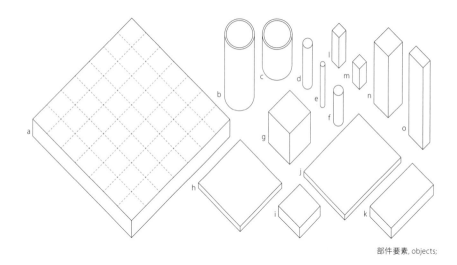

部件要素, objects;

部件要素列表（单位为一个模数）

a. 底座，21见方，厚2.5，上表面刻有7等分的网格
b. 空心圆柱，高9，外直径为4.5，内直径为4
c. 空心圆柱，高4.5，外直径为4.5，内直径为4
d. 圆柱，直径为1.5，高6
e. 圆柱，直径为0.75，高6
f. 圆柱，直径为1.5，高4.5
g. 立方体，各边长为4.5
h. 平板，9×9，厚0.75
i. 平板，4.5×4.5，厚1.5
j. 平板，9×12，厚0.75
k. 平板，4.5×9，厚1.5
l. 长方体，1.5×1.5×4.5
m. 长方体，1.5×1.5×3
n. 长方体，3×3×9
o. 长方体，1.5×3×12

LIST OF OBJECTS (1 module as the unit)

a. base, 21 × 21 × 2.5, One side covered with 7 × 7 grids
b. 4.5 outside dia. × 9 hollowed cylinder, 4 inside dia.
c. 4.5 outside dia. × 4.5 hollowed cylinder, 4 inside dia.
d. 1.5 dia. × 6 column
e. 0.75 dia. × 6 column
f. 1.5 dia. × 4.5 column
g. 4.5 × 4.5 × 4.5 cube
h. 9 × 9 × 0.75 plate
i. 4.5 × 4.5 × 1.5 plate
j. 9 × 12 × 0.75 plate
k. 4.5 × 9 × 1.5 plate
l. 1.5 × 1.5 × 4.5 prism
m. 1.5 × 1.5 × 3 prism
n. 3 × 3 × 9 prism
o. 1.5 × 3 × 12 prism

小结 reflection

教学现场, in-class practice;　　　绘制轴测图, construct axonometric view;

平面图、立面图与剖面图是三种基本的正投形图，建筑师使用它们来想象、创造以及组织三维形式。一块有划分网格的底板，加上一组模数化的多面体物件（木质的体块、板片及圆柱体），是学习这类平面性图示的一种有效工具。练习之初，选择要素并将其布置于底板上，使各要素与底板上的网格线、节点对位。学生通过观察便会意识到，所有的要素在尺寸上均与底板网格的模数相配合，这一发现将有助于平面图的绘制。举例说明，比例为1:1:3的长方体，其方形截面各边是网格模数的一半，高度便是网格模数的1.5倍。

为帮助学生在制图中准确控制比例，A3绘图纸是预先设计的。其上排布有辅助绘制平面与立面的网格，其尺寸为底板网格的1:4。学生首先使用很轻的徒手线条在绘图纸上按要求勾画出平面、剖面与立面，然后再在这个草稿的基础上，使用正确的线宽线型，绘制高品质的徒手线条以完成图纸。

这个物件绘制练习除了提供绘图能力训练外，它还能够向学生介绍设计的相关形式属性，包括空间限定、对称与非对称、要素与场地（模型底板）之间的关系，以及最重要的"暗示的视觉连续性"概念。比

Plan, section, and elevation are the fundamental orthographic drawing types that architects use to visualize, create and articulate three-dimensional form. A useful device for learning the basic conventions of these *planimetric* drawing types is a physical model with gridded base and a set of modular prismatic objects in the form of wood blocks, plates and cylinders. At the beginning of the exercise a selection of elements is carefully arranged on the base in alignment with grid lines and nodes. This facilitates drawing the plan once it is observed that all elements are dimensioned modularly in coordination with the dimension of the grid. For example, a square prism of 1:1:3 proportions has a square cross section of a 1/2 grid module and its height is 1 1/2 grid modules.

To aid the student in constructing the drawings with correct proportions, a pre-formatted A3 drawing sheet of paper with plan and elevation grids matching (at a scale of 1:4) the model set-up base is provided. The student then lightly constructs the plan, section or elevation as required by freehand drawing and after, draws again in freehand the final lines with good line quality and correct weight over the construction lines.

Aside from providing a drawing lesson, the object set-up drawing exercise also has the capability to introduce *formal* aspects of design, such as spatial definition, symmetry versus asymmetry, relationship of elements to

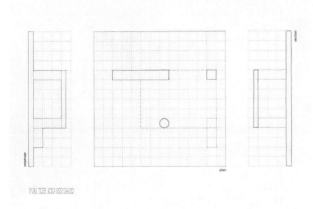

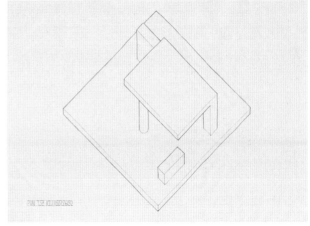

学生作业, student work (PUN, Tsz Kiu);

如说，通过对位布置两个或更多的物件，可以获得一个具有固定高度与确定长度的垂直表面，而这些物件之间可能存在着间隔或不连续的部分。诸如此类的设计关系可能会在课程训练过程中由指导教师指出并展开讨论。

练习的第二部分涉及绘制物件模型的轴测图。使用前一阶段完成的平面图，将其按照一定角度（如45°）旋转后置于黄色草图纸下作为衬底，然后以此描线并徒手建构轴测图。首先应草图勾勒一下，然后从图面的下部（模型相对靠前的部分）开始，确定物件之间的相互遮挡，以减少在最终图纸中因为线条可能造成的误解。轴测图中，各垂直线条的高度从立面图或剖面图中获得。图中所有线条均应按照轮廓线绘制，无需表达被遮挡的线条。

这一过程向学生介绍了一项值得频繁重复提及的重要技巧：我们可以将草图纸作为一个工具，用它来描摹、发展一层层的图纸，设计师可以在这个删减、增加的过程中不断对设计想法进行变形。对于草拟、构建衬底用的第一稿草图，使用相对便宜且半透明的绘图纸便是权宜之计。

the field (model base) and most important, the concept of *implied visual continuity*. For example, a vertical surface of a fixed height and certain length may be implied by two or more objects in alignment despite a gap or other discontinuity existing. These relationships and others may be pointed out and discussed by the instructor in the course of the exercise.

A second part of the exercise involves the construction of an axonometric drawing of the model set-up. Using the completed plan drawing from part one the student is asked to rotate the plan to a preferred angle (e.g. 45°) and then, using yellow tracing paper as an overlay, trace and construct the axon by freehand. The drawing should be visualized first and then begun in the lower (and front) view so as to minimize the confusion of which lines in the completed axon are "hidden" behind the view of the object that can be seen. The height of vertical lines in the axon is taken from the elevation or section drawings. All lines should be drawn as profile lines with no hidden lines shown.

This process introduces an important skill that should be reiterated frequently: tracing paper is a tool for tracing and developing layers of drawings that allows the designer to *transform* an idea through deletion and addition. Inexpensive and opaque drawing paper is more expedient for sketching and constructing underlay first drafts.

3.2 形式 [要素] 练习
exercises for **form [elements]**

3.2.1
实体 mass

任务书 handout

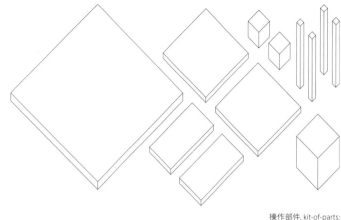

"建筑，就是在光的环境中将实体体量正确、熟练、有序地组织在一起。"——勒·柯布西耶

"Architecture is the correct, masterful, and orderly play of masses brought together in light." - Le Corbusier

训练要求

空间与体量可能被认为是一组相对概念。一个虚空体积可以认为是空间，一个实体体积可以认为是体量。练习的核心在于形式要素的组合操作，通过研究不同形式要素的形状、摆布位置及彼此关系来创造一个单一体量。体量的限定与接合是这个设计的关键，体量的形式应该是清晰可读的。为表达实体体量的主要特征，便要着重关注要素间的积聚。若将要素松散排布，就不可能产生整体而结实的效果。

练习中所使用的要素部件（如立方体，棱柱等）均有其自身形式的强烈特征。将它们组合在一起，消解要素部件各自的独立性，并生成一个具有整体性的构成是这个练习的成败关键。

在练习中，使用下述提供的要素部件进行设计。模型底板大小为144mm x 144mm。最终模型应使用木工胶粘实固定，使用精细的砂纸对其表面进行打磨处理，最后刷上白色石膏。

REQUIREMENTS

Space and mass may be thought of as opposites. A volumetric void may be thought of as a space; a volumetric solid as a mass. The goal of this exercise is to create an arrangement of elemental forms that, by their shape, position, and relationship to each other, define a single mass. The definition and articulation of this mass is to be the central idea of the design. It should be perceptually clear and unambiguous; in other words, legible. The primary characteristic of mass, solidity should not be compromised by a loose arrangement of the elements.

In creating a composition, which employs separate object elements (cube, rod, etc.), it is important to recognize that these objects have a natural tendency to remain independent and freestanding. Overcoming this tendency towards separateness is a key to creating a unified composition.

Create a design using the given elements described below. The field is a 144 mm x 144 mm base. The final model should use the wood elements provided and permanently fastened with wood glue. The model will be finished in white gesso.

石膏, 海绵刷, 笔刷, acrylic white gesso, foam brush, painting brush;

1个立方体	36mm x 36mm x 36mm		1 cube	36mm x 36mm x 36mm
2个小立方体	18mm x 18mm x 18mm		2 prisms	18mm x 18mm x 18mm
2片板片	72mm x 72mm x 6mm		2 planes	72mm x 72mm x 6mm
2片板片	72mm x 36mm x 6mm		2 planes	72mm x 36mm x 6mm
4根棱柱	72mm x 6mm x 6mm		4 rods	72mm x 6mm x 6mm
1块底板	144mm x 144mm x 9mm		1 base	144mm x 144mm x 9mm

评价标准

- 满足训练目标
- 设计想法及构成的清晰与品质
- 部件之间、部件与场地的关系

绘图记录

在一张水平放置的A3绘图纸上，等比例（1:1）绘制一个平面图、两个剖面及一个立面图。在另一张A3绘图纸上绘制倾角为45°的轴测图（亦可采用其他角度，以清晰展现体量形式为目的）。所有的图均应布置在绘图纸的中央。绘图时应合理选择线宽。提交成图前核查错误或遗漏的线条。

EVALUATION CRITERIA

- Compliance with program objectives
- Clarity and quality of the idea and composition
- Relationship of elements to each other and to the field

DRAWING DOCUMENTATION

On one A3 sheet (horizontal orientation), draw the plan, two sections and an elevation at the scale of 1:1 (full scale). On a second A3 sheet (vertical orientation), draw an axonometric at 45° inclination. Another angle may be selected if such angle can better present the design. These drawings should be centered on the paper. Be sure to use good line weight. Check drawings for mistakes or missing lines before handing in.

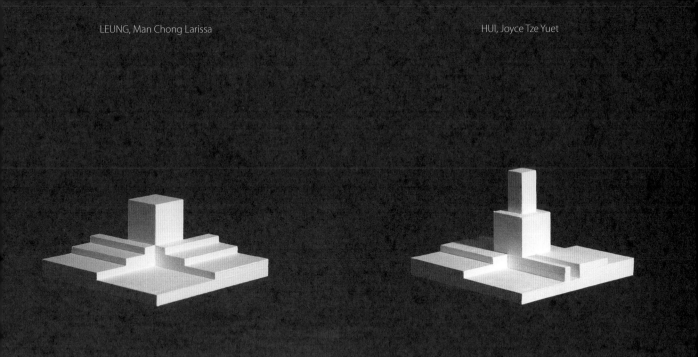

LEUNG, Man Chong Larissa

HUI, Joyce Tze Yuet

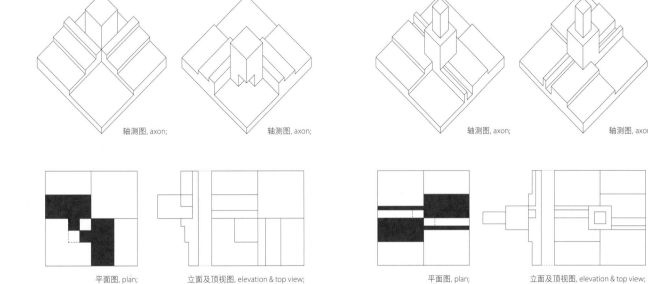

轴测图, axon;

轴测图, axon;

轴测图, axon;

轴测图, axon;

平面图, plan;

立面及顶视图, elevation & top view;

平面图, plan;

立面及顶视图, elevation & top view;

83

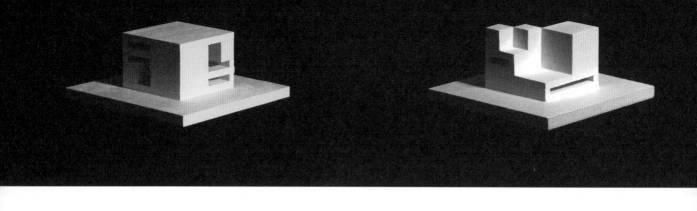

轴测图, axon;

轴测图, axon;

轴测图, axon;

轴测图, axon;

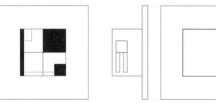

平面图, plan;

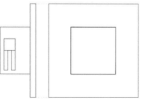

立面及顶视图, elevation & top view;

平面图, plan;

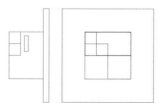

立面及顶视图, elevation & top view;

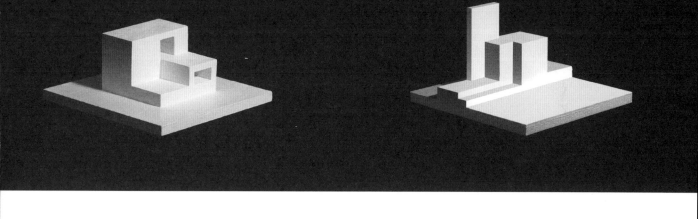

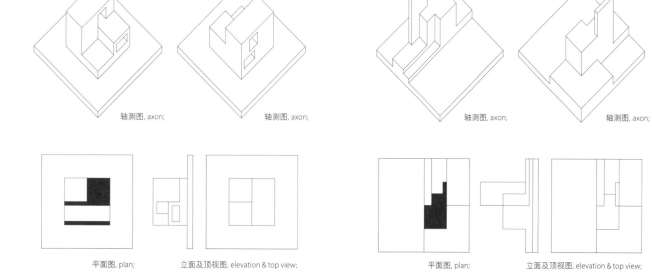

轴测图, axon;　　　　轴测图, axon;　　　　轴测图, axon;　　　　轴测图, axon;

平面图, plan;　　　　立面及顶视图, elevation & top view;　　　　平面图, plan;　　　　立面及顶视图, elevation & top view;

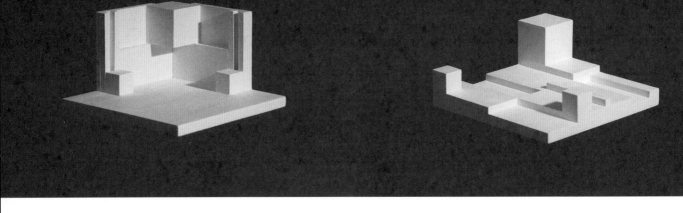

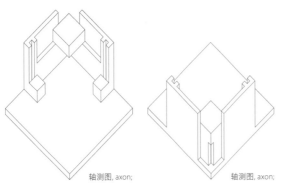

轴测图, axon;　　　　　　　　轴测图, axon;

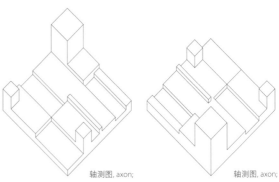

轴测图, axon;　　　　　　　　轴测图, axon;

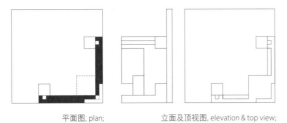

平面图, plan;　　　　立面及顶视图, elevation & top view;

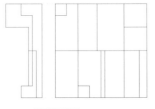

平面图, plan;　　　　立面及顶视图, elevation & top view;

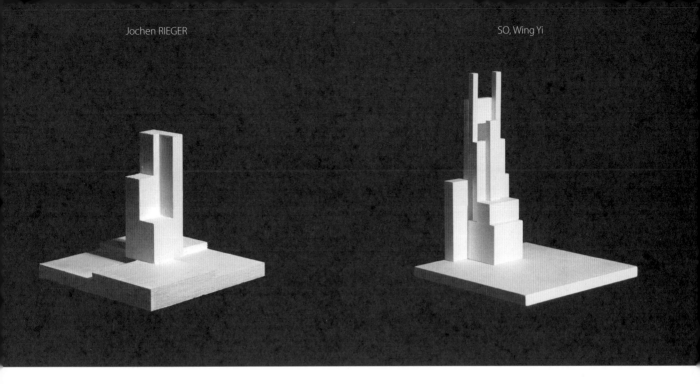

Jochen RIEGER

SO, Wing Yi

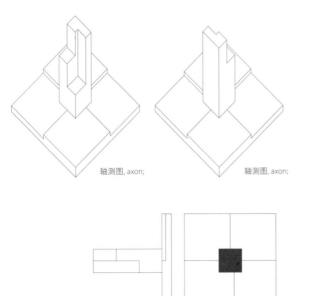

轴测图, axon;

轴测图, axon;

立面及平面图, elevation & plan;

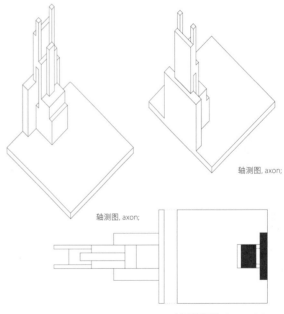

轴测图, axon;

轴测图, axon;

轴测图, axon;

立面及平面图, elevation & plan;

87

3.2.2

容积 volume

任务书 handout

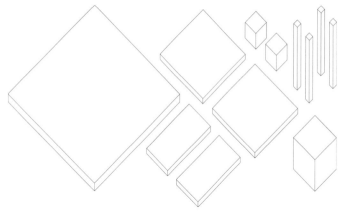

操作部件, kit-of-parts;

"如果要我用简单的词语来定义建筑，我会说，建筑是慎思的空间制造。"——路易斯·康

"If I were to define architecture in a word, I would say that architecture is the thoughtful making of spaces." - Louis Kahn

训练要求

练习的核心在于形式要素的组合操作，通过研究不同形式要素的形状、摆布位置及彼此关系来创造一个空间体量。空间的限定与接合是这个设计的关键，其形式应该是清晰可读、界定明确的。诸如空间特点、围合程度、几何属性、朝向等因素是不明确的，练习中学生应就此展开探索，使用给定的要素部件（如立方体，棱柱等）来塑造一个空间。需要注意的是，各要素部件均有其自身形式的强烈特征。将它们组合在一起，消解要素部件各自的独立性并生成一个统一的整体，是这个练习的成败关键。

在练习中，使用下述提供的要素部件进行设计。模型底板大小为144mm x 144mm。最终模型应使用木工胶粘实固定，使用精细的砂纸对其表面进行打磨处理，最后刷上白色石膏。

1个立方体　　36mm x 36mm x 36mm
2个小立方体　18mm x 18mm x 18mm

REQUIREMENTS

The goal of this exercise is to create an arrangement of forms that, by their shape, position, and relationship to each other, define a space, a single volume. The definition and articulation of this space is to be the central idea of the design. It must be perceptually clear and unambiguous; in other words, legible. The characteristics of the space; degree of closure, geometry, orientation, etc., are unspecified and are for you to explore. In creating a composition that employs object elements (cube, rod, etc.) to define a space, it is important to recognize that these objects have a natural tendency to remain independent and freestanding. Overcoming this tendency towards separateness is a key to creating a unified composition.

Create a design using the given elements described below. The field is a 144 mm x 144 mm base. The final model should use the wood elements provided and permanently fastened with wood glue. The model will be finished in white gesso.

1 cube　　　36mm x 36mm x 36mm
2 prisms　　18mm x 18mm x 18mm

89

2片板片	72mm x 72mm x 6mm
2片板片	72mm x 36mm x 6mm
4根棱柱	72mm x 6mm x 6mm
1块底板	144mm x 144mm x 9mm

2 planes	72mm x 72mm x 6mm
2 planes	72mm x 36mm x 6mm
4 rods	72mm x 6mm x 6mm
1 base	144mm x 144mm x 9mm

评价标准

– 满足训练目标
– 设计想法及构成的清晰与品质
– 部件之间、部件与场地的关系

绘图记录

　　在一张水平放置的A3绘图纸上，等比例（1:1）绘制一个平面图、两个剖面及一个立面图。在另一张A3绘图纸上绘制倾角为45°的轴测图（亦可采用其他角度，以清晰展现体量形式为目的）。所有的图均应布置在绘图纸的中央。绘图时应合理选择线宽。提交成图前核查错误，或遗漏的线条。

EVALUATION CRITERIA

- Compliance with program objectives;
- Clarity and quality of the idea and composition;
- Relationship of elements to each other and to the field.

DRAWING DOCUMENTATION

On one A3 sheet (horizontal orientation), draw the plan, two sections and an elevation at the scale of 1:1 (full scale). On a second A3 sheet (vertical orientation), draw an axonometric at 45° inclination. Another angle may be selected if such angle can better present the design. These drawings should be centered on the paper. Be sure to use good line weight. Check drawings for mistakes or missing lines before handing in.

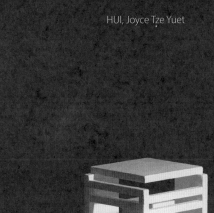

HUI, Joyce Tze Yuet

LI, Ho Kong Jeremy

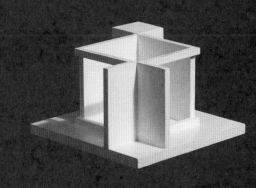

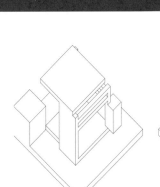

轴测图, axon;　　　　轴测图, axon;

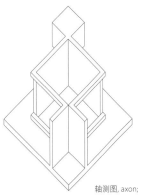

轴测图, axon;　　　　轴测图, axon;

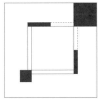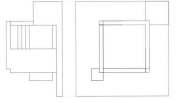

平面图, plan;　　　立面及顶视图, elevation & top view;

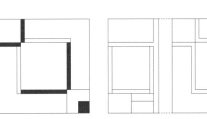

平面图, plan;　　　立面及顶视图, elevation & top view;

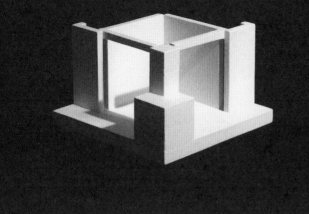
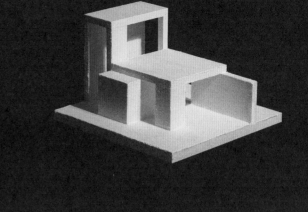

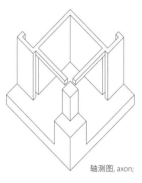

轴测图, axon;

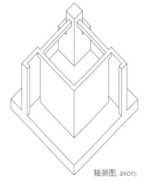

轴测图, axon;

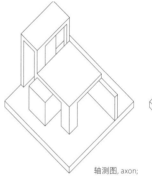

轴测图, axon;

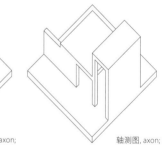

轴测图, axon;

平面图, plan;

立面及顶视图, elevation & top view;

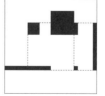

平面图, plan;

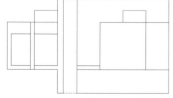

立面及顶视图, elevation & top view;

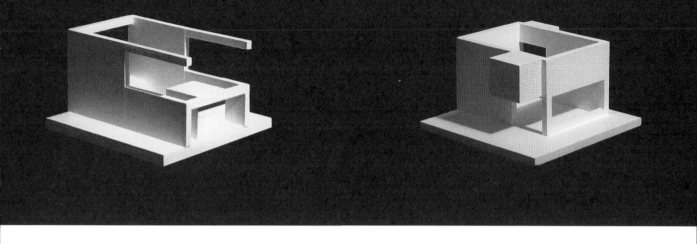

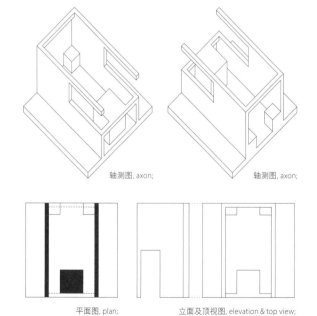

轴测图, axon;　　　　　轴测图, axon;

平面图, plan;　　　　立面及顶视图, elevation & top view;

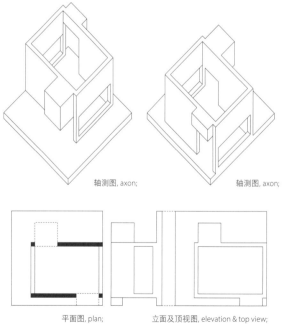

轴测图, axon;　　　　　轴测图, axon;

平面图, plan;　　　　立面及顶视图, elevation & top view;

93

CHEUNG, Ming Chung

TSE, Yan Hei Justin

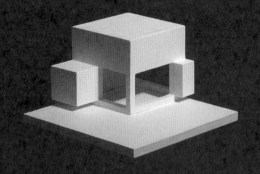

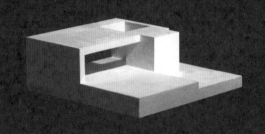

轴测图, axon;

轴测图, axon;

轴测图, axon;

轴测图, axon;

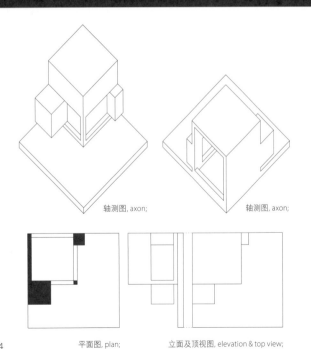

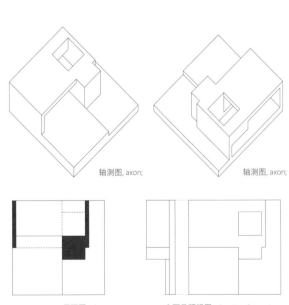

平面图, plan;

立面及顶视图, elevation & top view;

平面图, plan;

立面及顶视图, elevation & top view;

WONG, Hoi Tung Cheryl

SO, Wing Yi

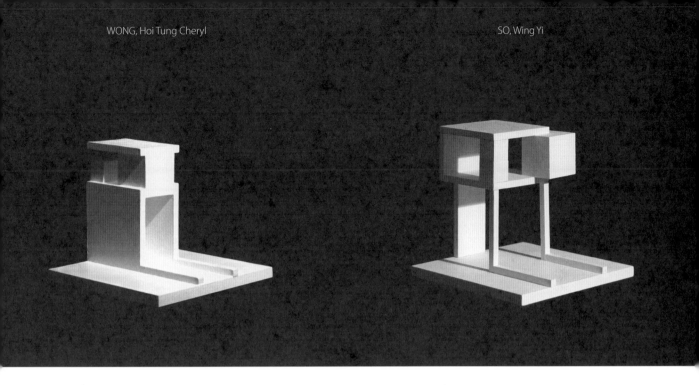

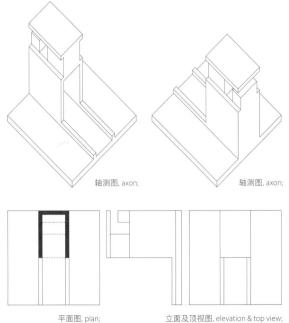

轴测图, axon;　　　　　　　　　　　　　　轴测图, axon;

平面图, plan;　　　　　　　立面及顶视图, elevation & top view;

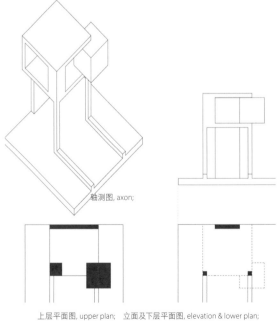

轴测图, axon;

上层平面图, upper plan;　　立面及下层平面图, elevation & lower plan;

小结 reflection

研究模型, work on study models;

　　"实体"与"容积"这两个练习所处理的是关于形式的想法。由于感知物体是一种与生俱来的能力，初学设计的学生们通常会比较熟悉实体形式。然而，捕捉空间形式的能力往往需要经过训练培养。"容积"练习要求学生塑造一个形状、尺寸均明确可读的空间容积。限定空间容积离不开实体部件要素的组织，由此生成的容积的三维形式不可以是模棱两可的。所以，对于学生而言，明确空间限定的原则是十分必要的，空间中的8个点可以确定一个立方体形式的各边角，并以此准确确定其形状及尺寸。组合使用点、线、面可取得一个相似的结果。

　　在这两个练习中都涉及关于实体与虚空的概念。学生可以通过一系列的加法、减法操作来生成一个复杂的实体形式构成。加法操作指的是将不同的要素组合在一起，以获得一个复杂的体块，这比减法操作更容易理解。通过加法的形式创造，独立的、形状不同的实体可以组合成一个复杂的形式，这个形式是直观、具体的。减法的形式创造则要求对虚空进行感知，移动实体的要素来限定虚空。"实体"练习中所强调的"减法"操作是指在一个实体体量中存在一些可

The two exercises, Mass and Volume, address the idea of form. Beginning design students are generally familiar with the form of solids since the perception of objects is innate. The visualization of spatial form, however, requires practice. The exercise Volume requires the forming of a single spatial volume whose shape and dimension is explicit and legible. This requires that the solids defining the volume are arranged in such a way that there is no ambiguity in the three-dimensional form of the volume. The principles of space definition apply here. Eight points in space can determine the corners of a cubic form thereby establishing its shape and dimension with precision. Combinations of points, lines and planes can achieve a similar result.

Imbedded in both exercises is the notion of solid and void. The composition of a complex solid form can be achieved through either the process of *addition* or *subtraction*. Additive form is obtained by the assemblage of various elements to create a complex object and is more easily understood than the process of subtraction. With additive form the process of combining separate solid, shaped elements into a complex form is tangible and visible. Subtractive form making requires, however, perception of the void and the removal of solids in defining the void. In the exercise Mass, the process of subtraction is emphasized. The underlying objective is to define a single compact mass that can be visualized as a form with subtractive voids. The

制作最终模型, paint the final model; 图纸记录, drawing documentation; 评图现场, project review;

以感知为挖去的虚空部分，这是训练的一个潜在目的。实体体量应当感知为一个整体，其中虚空部分的形状以及它们与整体形式之间的关系应当成为构成中独立的要素。而在"容积"练习中，设计的过程更接近"加法"操作。利用部件要素的表面、边缘及角落来限定虚空形式，赋予该容积以可读的形式，这是练习的基本目标。

两个练习所使用的部件要素是完全相同的。从实际的角度来看，这会简化练习的准备工作。而站在教学法的角度，有意训练学生从相同的要素出发，创造出本质不同的设计。部件要素之间的模数关系对于要素的组合以及构成的品质是至关重要的。练习使用的底板也可以看作是一件部件要素，由此建立起具有模数尺寸的方形场地，激发各物件与底板之间准确的形式关系。意识到这一点是十分关键的，这样底板就不再被当作是物件摆放的背景，或沦落为一个无关的表面。在"实体"练习中，要素之间的模数关系可结合底板发展出一个被深深侵蚀的地表，一个由实体延伸出的景观。它所创造出的构成将底板（或场地）与实体物件强烈地统一在一起。

configuration of the mass object as a whole, the shape of the voids and their relationship to the overall form are interdependent elements of the composition. In the exercise Volume the approach tends to be more additive. The surfaces, edges and corners of the kit of parts elements must be arranged to define the shape of a void; the single volume whose legible form is the principal goal.

The kit of parts for Mass and Volume is identical. From a practical standpoint this simplifies the preparation of the materials for both exercises. Pedagogically there is an intention to demonstrate that identical elements can generate fundamentally different designs. The elements of the kit of parts are modular. This is a critical feature that contributes significantly to the ease of combination of the elements and the quality of the composition. The base is seen as an element itself. It establishes a square field of modular dimension encouraging precise formal relationships between the object and the base. This is important so that the base is not regarded as an unrelated surface seen purely as a backdrop for the object. In the exercise Mass, the modularity of the elements can be used in conjunction with the base to develop a deep eroded ground surface, a landscape that is designed as an extension of the mass. This creates a composition that strongly unifies the base or field with the mass object.

97

3.3 构成［空间］练习
exercises for composition [space]

3.3.1
图底关系 figure and ground

任务书 handout

72 mm x 72 mm

108 mm x 54 mm

144 mm x 36 mm

216 mm x 24 mm

252 mm x 18 mm

黑色矩形要素, black rectangular elements;

正方形这一形状看似中性，实则暗含了许多激发形式生成的力量。直角与对边平行暗示了其正交网格划分的可能，应用不同的模数与节奏，如田字格、九宫格，或是A-B-A组合的"格纹"。正方形的对角还存在均衡的力，暗示斜向上的动态。此外，在正方形中可绘制圆，使其与方形各边的中点相内切，进一步强化中心与边缘的特征。

正方形的这些特征，可以为创造一个整体、同一的构成设计提供最初的思路方向，并以此强化场地的内在几何性质。练习中，仅使用给定的黑色长方形纸片，将它们摆放在白色底板上进行构图组织：可以将黑色纸片作为图，而白色底板部分阅读为底；或者反过来以纸片间的白色部分为图，将黑色部分阅读为底。这一现象可以称为图底反转。

训练要求

在252mm x 252mm的白色底板上，使用规定尺寸的一部分黑色纸片进行构图组织。并在这一过程中，探索正方形的几何特征以及"图底模棱两可"这一概念。练习中使用的纸片要素尺寸如上图所列。学生可

The seemingly neutral shape of a square embodies many forces that can generate form. The right angle and parallel sides imply an orthogonal grid capable of subdivision into various modules and rhythms such as 4-square, 9-square, or A-B-A "tartan plaids". There is also equal strength of opposite corners, which imply diagonal flow. Furthermore, a circle can be inscribed within the square, with tangents touching the midpoint of each side, reinforcing the property of centrality and rest.

These properties of the square may provide an initial direction in creating a unified composition that reinforces the inherent geometric properties of the field. Using only black rectangular elements on a white field a composition may be developed that establishes either black elements as "figure" with the white areas reading as ground, or conversely, white areas emerging as figure with the black areas reading as ground. This phenomenon is known as figure/ground reversibility.

REQUIREMENTS

On a 252mm x 252mm. Using rectangles of given size cut from black paper, create a composition that explores the properties of the square and exploits the idea "figure/ground ambiguity". For the composition, use only the dimensioned elements listed above. You are not required to use all of

根据需要从中选取，不要求使用全部的要素类型，某些要素亦可重复多次使用。矩形纸片之间相互不可交叠、可边缘相接。确保纸片平整地粘贴于白色底板上，表面无气泡。

评价标准

- 图底模棱两可的品质
- 对方形几何特征的探索
- 要素与场地之间的关系
- 对比、层级与比例
- 模型制作的准确与品质

绘图记录

对设计模型进行三维投形。对模型中黑白部分之间的图底关系进行解读，赋予其中白色或黑色的部分以高度。升起部分的体量高度可以是18mm，24mm或36mm。在这一过程中，应尝试通过三维体量进一步强化原二维模型中的图底关系阅读。最后，将这一三维体量以轴测图的方式记录于A3绘图纸上，按照1:2的比例徒手绘制线条。

the elements and you may use some elements more than once. The rectangular elements cannot overlap but they can be touching. Be sure to glue the surfaces evenly in order to avoid bumps.

EVALUATION CRITERIA

- Quality of figure/ground ambiguity
- Exploration of the properties of the square
- Relationship of the elements to the field
- Contrast, hierarchy, proportion
- Precision and quality of construction

DRAWING DOCUMENTATION

Invent a three dimensional projection for the design. Interpret the black and white figure/ground of the exercise as a volumetric relief. The height of any portion of either the black area or the white area (but not both) can be drawn at a height of 18mm, 24mm or 36mm. Attempt to use the third-dimension to reinforce or strengthen figures or field readings that exist in the two-dimensional study. One axonometric projection describing the envisioned relief. Freehand pencil drawing on A3 drawing paper with 1:2 scale.

CHAN, Tsz Him

SO, Ling Sum Evangeline

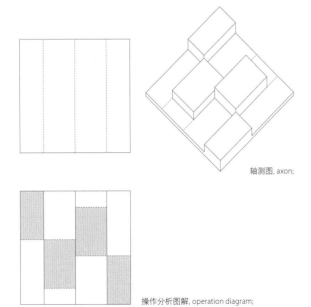

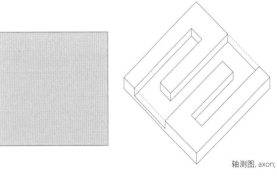

轴测图, axon;

操作分析图解, operation diagram;

轴测图, axon;

操作分析图解, operation diagram;

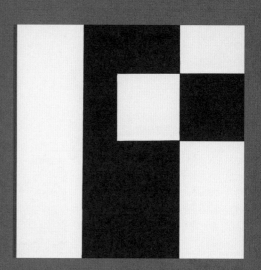

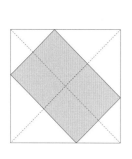

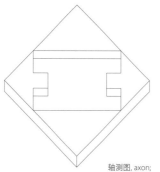

轴测图, axon;

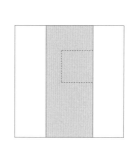

轴测图, axon;

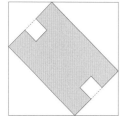

操作分析图解, operation diagram;

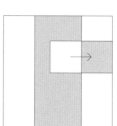

操作分析图解, operation diagram;

LO, Hoi Yau

WONG, Yik Chun

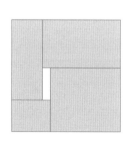

轴测图, axon;

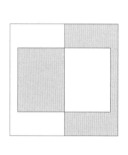

轴测图, axon;

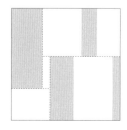

操作分析图解, operation diagram;

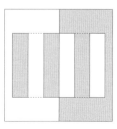

操作分析图解, operation diagram;

SUNG, Chen Ru

LEUNG, Kin Kan

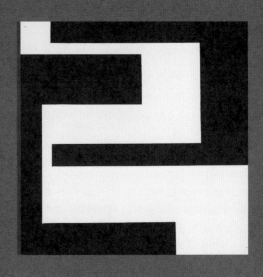

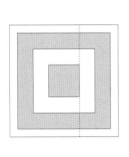

轴测图, axon;

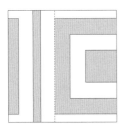

操作分析图解, operation diagram;

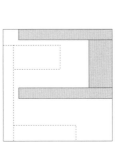

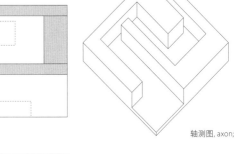

轴测图, axon;

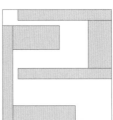

操作分析图解, operation diagram;

LIZHUANG, Minyi

AMADEO, Amelia

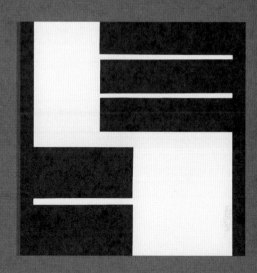

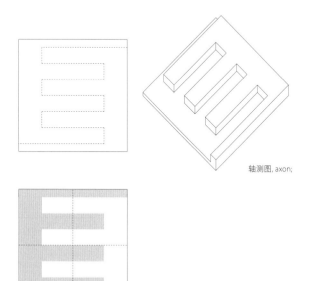

轴测图, axon;

操作分析图解, operation diagram;

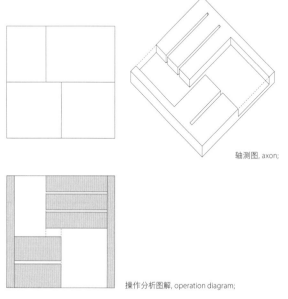

轴测图, axon;

操作分析图解, operation diagram;

变式 variation

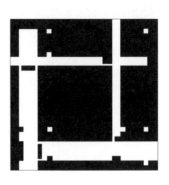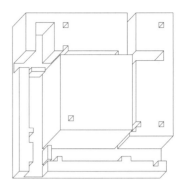

作业示例, example;

正方形这一形状看似中性，实则暗含了许多激发形式生成的力量。其形式操作的可能性可以同乐谱中音符与调式的组织相类比：如果两者的关系是有意义且可以被研究的，那么音乐可能就是我们期望的结果。

The seemingly neutral shape of a square embodies many forces that can generate form. And the possibilities might be seen as analogous to the arrangement and pattern of notes in a musical score: if the relationships are studied and meaningful, perhaps music will be the result.

训练要求

在252mm x 252mm、厚5mm的白色泡沫底板上，使用不同尺寸的黑色、白色方形纸片进行构图组织。并在这一过程中，探索正方形的几何特征以及"图底模棱两可"这一概念。最后，纸片需完全覆盖于底板表面上。在操作中，可以在底板上先覆盖一层黑色或白色的纸。应确保纸片平整地粘贴，表面无气泡。

对设计模型进行三维投形，通过三维体量来解读二维模型中黑白部分之间的图底关系。选择将其中白色或黑色部分升高，升起部分的体量高度不得超过36mm。在这一过程中，应尝试通过三维体量进一步强化原二维模型中的图底关系阅读。最后，将这一三维体量以轴测图的方式记录于A3绘图纸上，按照1:2的比例徒手绘制线条。

REQUIREMENTS

252mm x 252mm base of 5mm form board. Using squares of varying size cut from black and white paper, make a composition which explores the properties of the square and exploits the characteristics of the figure/ground concept. The entire surface of the base must be covered. This may be done with a first layer of either black or white paper. Be sure to glue the surface evenly in order to avoid bumps.

Invent a three-dimensional projection to interpret the black/white figure ground of the exercise as a volumetric relief. The height of any protion of the composition can be between 0 (left flat) and 36mm. Attempt to use the third dimension to reinforce or strengthen figures or field reading, which exist in the two-dimensional study. One axonometric projection describing the envisioned relief. Freehand pencil drawing on A3 drawing paper with 1:2 scale.

小结 reflection

图与底之间的差别以及它们之间可相互转换的特点，同实与虚的关系并无不同。体量与实体性关联，而体积与虚无或者空的状态有关，两者是相互对立的。但我们仍可以采用与同体量相同的形式特征来看待体积，包括形状，边界以及尺寸。

通常，我们倾向于将实体描述为"图形"。实体物体具有明确的形状，它的平面是某特定高度上的一条横剖切线，这条线表达了该物件的轮廓。它是一个闭合的图形，可以被阅读为图形。对这个形状的内部填色（如将其涂黑），可以赋予它更强烈的实体性，创造其本身与周围区域（通常所指的"底"）之间剧烈的视觉反差。

研究显示，人们天生具有感知图形的能力，而对于底的感知则相对较弱。我们会关注于那些围绕我们的实体存在，它们通常表现为物体的形式。而空间设计则要求我们对于底的感知高度敏感，也就是"余留空间"或"夹间空间"。图底关系练习主张通过图形设计来强化对于底的感知，它关注于图底关系的模棱两可。通过使用黑色的纸片来创造一个构成，其中黑色与白色部分均不是主导的图形。在构成的某一部

The distinction between figure and ground, and its reversibility, is not unlike the interchangeability of solid and void. Mass is associated with solidity while volume is the condition of emptiness or void, the opposite of solid. Yet we can ascribe to volume the same attributes of form as mass: shape, boundary and dimension.

Solids tend to be described as *figure*. Objects are solids and they have explicit shape. The plan of a solid is a section cut line made at a particular height. It is a line that describes the profile of the object, a shape that always has closure and is read as a figure. Filling in the interior of the shape (with the color black, for example) gives the figure more solidity, creating more visual contrast between itself and the surrounding area that we refer to as the *ground*.

Studies have shown that we are pre-conditioned to perceive figure over ground. Our tendency is to focus on the solid reality that surrounds us, generally represented as objects. Spatial design requires a heightened sensitivity towards the perception of ground, sometimes referred to as the *leftover space* or *in-between space*. Exercise 3.1 *figure and ground*, is a graphic design exercise intended to strengthen the perception of ground. It emphasizes the reversibility of the figure/ground. Using black cutout elements, a composition is made in which neither black nor white is dominant figure. In

分，白色会因其尺寸、位置或形状的特点而表现为图形，而在另一部分，黑色更具有图形的品质，这就在黑色与白色之间创造了一个变化的平衡。到某一程度时，两者之间共同作用的本质会得到凸显，进而强调实体与虚空之间的可互换性。

为强化对这一点的认识，练习要求将图底构成中黑色或白色部分择其一，然后沿垂直方向升起，以实现三维的再现。现在，构成中图形的部分升起成为实体，底的部分表现为虚空，两者之间的对比所暗示的是一个建筑设计（也可能是城市设计）的图示。练习直接控制的是黑色与白色的形状，以此来实现图底之间的模棱两可，由此生成的三维实体（物体）与虚空（空间）将同时具备图形的品质。其间的空间部分同样具有可读的形式，它是整体构图中不可或缺的一部分。

在练习的变式中，我们删去了仅可以使用具有模数尺寸关系的纸片的限制条件，学生在塑造黑色部分时拥有更大的自由度。变式练习中，唯一的限制条件是：构图中黑色或白色区域只能由矩形构成。需要注意的是，设计操作同样可以是从黑色的背景中挖去一些白色的形状。练习同样需要基于图底构成来创造一

some areas, by virtue of size/position/shape characteristics white emerges as figure. In another section black will tend to acquire figural qualities. This can lead to a fluctuating balance between black and white that demonstrates to an extent, their reciprocal nature, and hence, the reversibility of solid and void.

To accentuate this aspect, a three-dimensional representation of the figure-ground composition is made by extruding either black or white areas in the vertical dimension. The contrast between the extruded figures now perceived as solids, and the ground which now represents the voids, is suggestive of an architectural or perhaps, urban design scheme. The effort to control both the black and the white in the attempt to create ambiguity between figure and ground has led to a result in which both the solid (object) and the void (space) have figural quality. The space in between has acquired legible form and has been made an integral part of the overall composition.

A variation to this exercise removes the restriction of using only the modular dimensioned rectangles in favor of more freedom in shaping the black areas. The only restriction is that all black (or white) areas of the composition be created with square or rectangular shapes. Note that one could just as well begin with white cutout shapes placed on a black back-

度量尺寸, measurement;

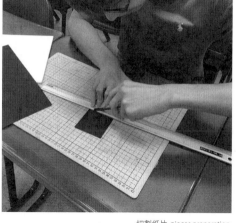

切割纸片, pieces preparation;

个立体浮雕，包括底板在内，总共层数不超过4层。由此升起的方式可以强化构成设计的图示或配置，或提供一种不同的图示解读方式。

ground. Again a three-dimensional relief based on the figure/ground composition and having no more than four levels (including the base) can produce an extrusion that reinforces the pattern or figuration of the design, or offer a different interpretation of the pattern.

3.3.2
空间限定 space definition

任务书 handout

在正方形场地上，使用高25mm的板片来限定4个矩形空间，其中有一个空间处于支配性地位。注意，设计过程中势必会产生一些细小的空间，它们不应对主要空间的感知产生干扰。这四个主要空间之间可能存在一定的序列关系。在处理空间与场地的关系时，可应用基本的组织原理，如轴线、对称、层级、基准、重复等。

训练要求

首先制作一块200mm × 200mm的模型底板，由5mm厚的泡沫板上覆约1.5mm厚的瓦楞纸板制成。然后在底板上布置25片高25mm、厚1.5mm的瓦楞纸板，纸板的长度不限。所有的纸板均为矩形，其上不可开洞口，垂直粘贴在底板上并保证高度为25mm。限定出的四个空间必须为矩形。

评价标准

- 满足训练目标与要求
- 设计中使用整块场地
- 对于空间限定的理解与表达

On a square field, using only 25mm high planes create a design, in which there are four clearly defined rectangular spaces, one of which is dominant. Other spaces may be part of the design but should not compete with or confuse the perception of the four principle spaces. The four principal spaces may involve a progression or sequence. Organizational principles such as axis, symmetry, hierarchy, datum, repetition, etc., should inform the design and its relationship to the field.

REQUIREMENTS

One 200mm x 200mm field, consisting of corrugated cardboard (1.5mm in thickness) and foam core board (5mm thickness). Only 25 corrugated cardboard planes. Planes are to be 1.5mm thickness and 25mm in height. Length is unrestricted. Planes must be rectangular with no holes or notches and should be mounted perpendicular to the field and 25mm in height. All four defined spaces must be rectangles.

EVALUATION CRITERIA

- Compliance with program objectives and requirements.
- Use of entire field in the design.
- Demonstration of an understanding of spatial definition.

- 设计想法以及构成的品质
- 最终模型的手工品质

绘图记录

对设计模型进行绘图记录，将模型中限定的4个长方形空间转换为实体体量，在A3绘图纸上用徒手线条绘制1:1的轴测图，倾角为45°。对于空间限定而言，确定空间容积的是墙体的内表面部分，需以此确定各体量尺寸。同时准确表达体量形式与场地之间的关系。

然后，在另一张A3绘图纸上，徒手绘制三幅最终设计的平面分析图解，比例尺为1:2。这些图解应讨论下述问题：

- 实体与空间：对图底关系的理解。在设计中，紧密排布的板片是可以感知为实体。在图解中，通过排斜线以表示实体部分，即底的部分。

- 网格分割：区分场地划分的主次层级关系，并以网格的形式表示。如果网格划分中存在明确的比例关系，可标注说明。

- 组织原则：图解表示构成操作的组织原则，如基准、轴线、组群、几何关系等。

DRAWING DOCUMENTATION

Draw a 1:1 scale axonometric plan projection, interpreting the four rectangular spaces of the design as solid volumes, on A3 drawing paper. Each volume should have the dimensions of the space defined by the inside faces of the walls which describe that space. Show the volumetric forms in their correct relationship to the field.

Draw three analytical diagrams of the final design on a second A3 drawing paper. These diagrams are to be freehand drawings at 1:2 scale. Following issues should be addressed.

- Solid-void: A figure/ground interpretation. Walls closely spaced in the design may be conceived of as solid areas. Use diagonal hatching to graphically describe solid areas (ground).

- Grid subdivision: Identify the major and minor subdivisions of the field and represent them as a grid of lines. Identify proportional relationships if they exist.

- Organizational concept: Illustrate the principal ordering concept(s) that structure your composition (e.g. datum, axis, grouping, geometry, etc.)

实体与空间, solid-void;

轴测图, axon;

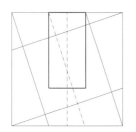

组织原则, ordering concept;

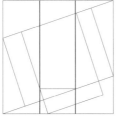

网格分割, grid subdivision;

实体与空间, solid-void;

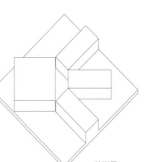

轴测图, axon;

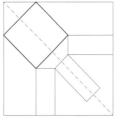

组织原则, ordering concept;

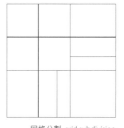

网格分割, grid subdivision;

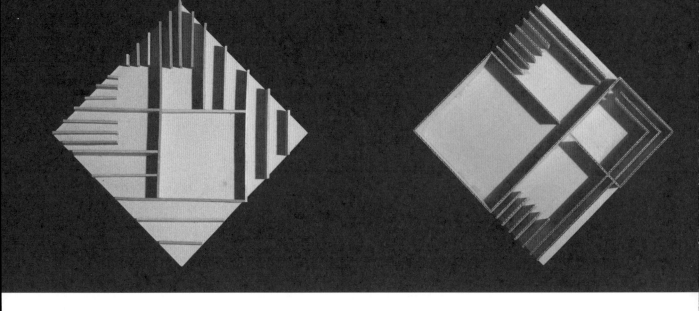

实体与空间, solid-void;

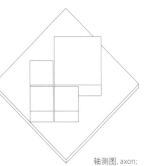

轴测图, axon;

组织原则, ordering concept;

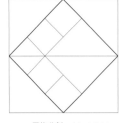

网格分割, grid subdivision;

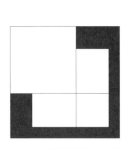

实体与空间, solid-void;

轴测图, axon;

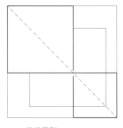

组织原则, ordering concept;

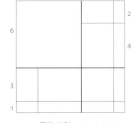

网格分割, grid subdivision;

IP, Kwok Hei Gloria

LAU, Cheuk Wang Michael

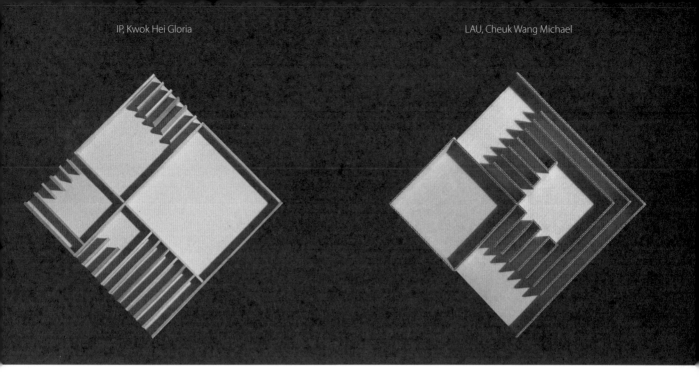

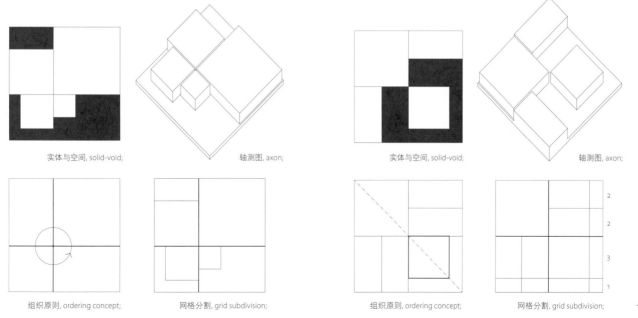

实体与空间, solid-void;

轴测图, axon;

实体与空间, solid-void;

轴测图, axon;

组织原则, ordering concept;

网格分割, grid subdivision;

组织原则, ordering concept;

网格分割, grid subdivision;

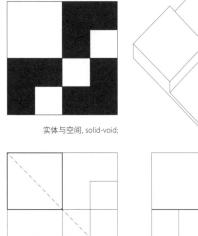

实体与空间, solid-void;

轴测图, axon;

实体与空间, solid-void;

轴测图, axon;

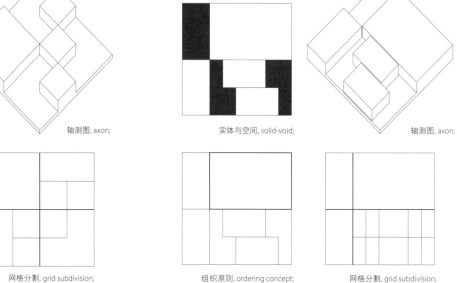

组织原则, ordering concept;

网格分割, grid subdivision;

组织原则, ordering concept;

网格分割, grid subdivision;

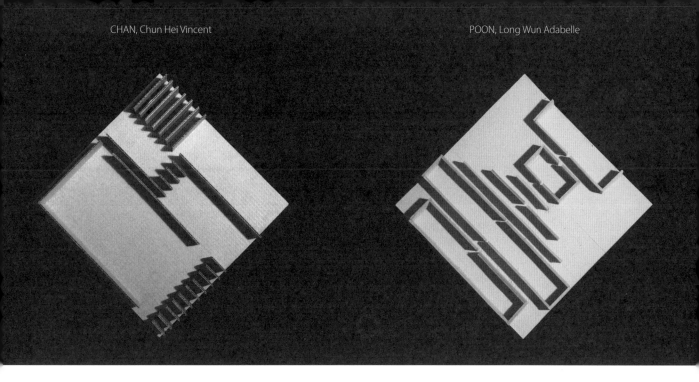

CHAN, Chun Hei Vincent

POON, Long Wun Adabelle

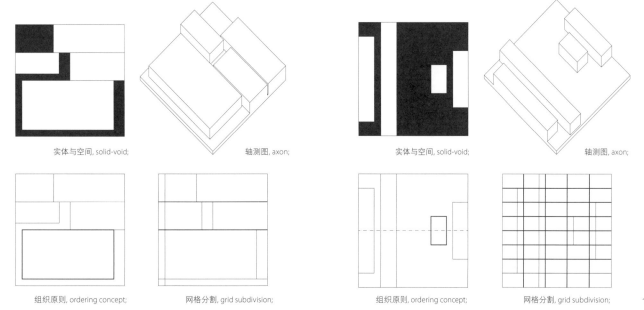

实体与空间, solid-void;

轴测图, axon;

实体与空间, solid-void;

轴测图, axon;

组织原则, ordering concept;

网格分割, grid subdivision;

组织原则, ordering concept;

网格分割, grid subdivision;

117

小结 reflection

制作过程模型, work on study models;

　　练习中涉及到一组适当的限制条件：空间需清楚限定，保持其矩形形状与可读性；设计中需完全使用方形场地；利用25片垂直板片来完成整个构成；创造出的4个空间需具备层级关系。一个成功的设计还可探索其他的关系与构成策略，如加法、减法空间生成之间的对比，空间序列组织。当垂直板片紧密排布时，可以创造出一定的密度，这使得从中移除部分板片时会产生一个挖去的空间。平行板片之间的开口可以表示流线或通道，由此创造出连通性——额外的一个设计要素。

　　完全利用方形底板，并不是要求将板片平均地布满整个场地。单一的片墙或是一组对位的墙端可以暗示设计对方形的划分。4个空间之间的模式与方形场地的划分组成了设计构思——构成的主要组织特点，它应当在设计中明确呈现。通常，缺乏一个清晰的设计构思，或是因组织策略失衡导致的繁复，会造成设计混乱而无法解决问题。设计的形式意义也因此无法得到清楚的传达。

　　在讲座"建筑的原理与要素"中列名的一些概念（轴线、对称、层级、基准与重复）可用以有效地组

This exercise has constraints without being overly restrictive. Space must be well defined, rectangular in shape, and legible. The field of the square must be wholly engaged. Twenty-five vertical planes must be used to complete the composition. Furthermore, *hierarchy* must be achieved in the relationship of four spaces. A successful scheme will explore other relationships and compositional strategies such as the contrast between additive and subtractive space making or the formation of a spatial sequence. Vertical planes may be closely spaced to imply density so that the removal of a portion of the planes will create a subtracted space. Openings between parallel planes can suggest circulation or path, and provide an additional element of connectivity.

To engage the entire field of the square base is not necessarily to distribute planes equally over the square. A single wall plane or the edges of many planes in alignment may imply subdivision of the square. The pattern of the four spaces, together with the subdivision of the square field will suggest a *parti*. This term refers to the dominant organizational character of the composition and should be evident in every design. Often a lack of clarity in the parti or the complication that arises from competing organizational strategies within the same design leads to confusion and lack of resolution. The formal meaning of the design is then not clearly expressed.

The concepts explained in the lecture *principles and elements of architec-*

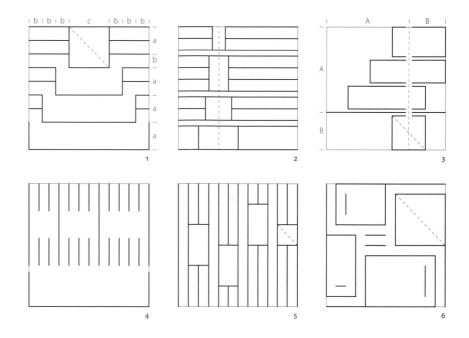

織設計。比如說，將4個空間的中心沿軸線組織，可以創造出一個對稱的布局 [圖1]。如這組對稱布置的空間與方形場地之間無對稱關係，它所創造的便是局部對稱 [圖2]。當軸線不經過各空間中心，而是對位於空間邊界上的開口，由此可創造聯貫各空間的流線，即縱貫 [圖3]。設計中創造層級性的方法多種多樣，尺寸差別是其中最明顯的一種。形狀（一個方形與三個矩形）或限定方式（利用牆體圍合空間邊界，加法；從密布的一組牆體中移除部分牆體，減法）是另外兩種策略 [圖4]。對於基準的應用相對較難，它可以是場地的一條邊界，4個空間沿著它對位布置。它還可以很微妙：在場地上均布一組平行牆，依其間距生成均質網格，然後於其中布置4個等寬不等長的矩形空間 [圖5]。最後一種生成秩序的原則是重復，可通過使用相似的模式或細節來建立群組內的聯繫 [圖6]。可以重復單一模式，或是就其尺度、比例進行微調。不過，這些形式可以閱讀為串聯的圖形群組，因為重復的緣故，群組中的各空間成為空間群中的一部分，它是實現統一的一種策略方法。

ture, namely axis, symmetry, hierarchy, datum and repetition, are useful in structuring a design for this exercise. The centers of the four spaces may, for example, be organized along an axis thereby creating a symmetrical arrangement [fig.1]. Symmetrical configurations of the spaces might be formed within the square field but not symmetric to the square, thus creating a local symmetry [fig.2].The axis might also be placed asymmetric to the centers of the spaces, aligned to portals in the walls of each space creating a circulation known as enfilade [fig.3]. Hierarchies can be suggested in many ways. Size is perhaps the most obvious, but shape (one square versus three rectangles) or definition (wall planes positioned to define the sides of the square: addition, versus wall planes removed from an area of closely spaced wall planes: subtraction) are two alternative strategies [fig.4]. The use of a datum is less well understood. A datum can be as simple as an edge along which the four spaces are aligned or it can be subtler. A grid of parallel wall planes of equal spacing across the square field of the base, in between which are inserted rectangular spaces of the same width but of varying length [fig.5]. The last ordering principle, repetition, can create group association through the use of a similar pattern or detail [fig.6]. The pattern may repeat identically or vary slightly in scale or proportion. Nevertheless, the forms read as concatenated figure groups and because of the repetition, the space associated with each group belongs to a family of spaces. This device acts as a unifying strategy.

3.3.3
基准墙 datum wall

任务书 handout

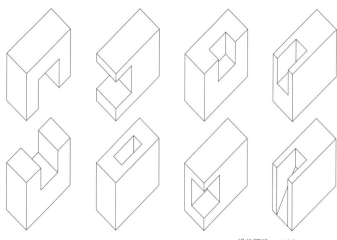

操作可能, possible operations;

步骤一

  在练习中，需创造一个墙体物件的形式，在此我们称之为"基准墙"，在之后它会作为形式的组织参照结构。由于这个墙体不具备任何具体的功能要求，所以对它的操作可以由自己的想法，仅需满足下述确定的一些要求。

  对于基准墙的变形是一个减法的挖去过程。在此可考虑的操作形式有两种：洞口或壁龛。洞口是指完全穿透墙厚的虚空，而壁龛则是挖去墙体的一角或是局部表面。通常壁龛不会完全穿透墙厚，否则便形如洞口一般。练习中，对于墙体的变形操作应控制在4次，如创造3个壁龛与1个洞口。就如何操作该墙体物件，可考虑以下原则：轴线，对称（局部对称），层级，比例，重复与基准。

  这个物件墙是一个长216mm、高36mm、厚15mm的长方体。对它的变形操作不能损失其本身的单体性，即保证它不被分割为多个独立体块。墙体由三层厚5mm的白色泡沫板叠在一起，你可以通过切割泡沫板来创造洞口和壁龛，但需要保持墙体的整体体量。洞口需紧贴底面（如门洞一般），高度为24mm，宽度

STEP 1

  In this exercise you are to create a form: a wall object. This element will later serve as an organizing reference structure we will call a datum wall. Since the wall does not have any specific functional requirement, it may be formed and articulated as you choose, within the parameters set forth below.

  The transformation of the wall object is a process of subtraction. Two types of operations can be performed on the wall object: holes and notches. A hole is a void passing through the thickness of the wall object. A notch is a subtraction on one of the edges or surfaces. It may penetrate the full thickness of the wall object (like a hole) or not. In transforming the wall object, you may make up to 4 operations total (e.g. 3 notches + 1 hole). In deciding how to transform the wall object consider the principles: axis, symmetry (local symmetry), hierarchy, proportion, repetition, and datum.

  The object wall is a rectangular prism of the following dimensions: 216mm x 36mm x15mm. It must be designed as an integral piece structurally, that is, connected together in a continuous manner. The object is made of three layers of white foam board (5mm thickness). You can cut out parts of the form board to create holes and notches, but maintain the given volume. The holes must extend to the base (like a portal) and be 24mm

121

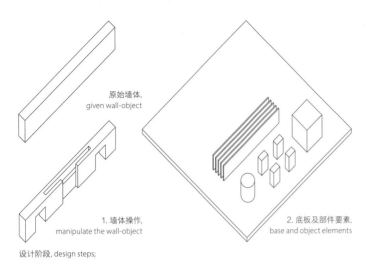

原始墙体,
given wall-object

1. 墙体操作,
manipulate the wall-object

2. 底板及部件要素,
base and object elements

设计阶段, design steps;

不得小于6mm或大于108mm。练习中不可创造弧形的洞口或壁龛。

high. The width of a hole should not be less than 6mm or greater than 108mm. Curved holes or notches are not allowed in this exercise.

步骤二

首先，在方形底板上放置前一步骤制作的基准墙。墙体摆放的位置会对整个场地的划分产生影响，应避免直接将其放置于底板的边缘。然后，使用下述给定的部件要素以及额外的一些垂直板片，来创造一个空间限定系统，并与基准墙建立秩序关系。重复、层级都是可考虑的组织策略。

模型底板大小为252mm x 252mm，通过在泡沫板上覆一层瓦楞纸板制作而成。垂直板片可使用厚度为1.5mm的瓦楞纸板，高度为24mm，长度不限。这些板片需垂直粘结在底板上，其上不可开挖洞口或缺口。部件要素如下：

1个立方体	36mm x 36mm x 36mm
1个圆柱体	24mm, 直径24mm
4个长方体	24mm x 15mm x 9mm

STEP2:

On a square field, carefully place the datum wall object from the previous step in such a way that it suggests a subdivision of the original square field (avoid placement of the wall object right against the edge). Using the elements provided below, plus a number of vertical planes, create a pattern of space definition that employs repetition, hierarchy and makes reference to the datum wall object.

One 252mm x 252mm field, consisting of a piece of a foam core board covered with a layer of corrugated cardboard. Vertical planes are to be 24mm in height, approximately 1.5mm thick and of any length (corrugated cardboard). They must be mounted in a vertical position. They should be solid with no notches or holes. The elements:

1 cube	36mm x 36mm x 36mm
1 cylinder	24mm x 24mmØ
4 prisms	24mm x 15mm x 9mm

评价标准
- 对于训练目标与要求的完成
- 设计中对场地的使用
- 对于"基准"的理解与表达
- 设计想法与构成的品质

绘图记录

在两张A3绘图纸上，使用徒手线条分别绘制1幅1:1的平面图及3幅1:2的平面分析图解，以作设计记录。分析图解可围绕下述问题展开：
- 场地上基准墙的摆放位置。
- 网格分割：区分场地划分的主次层级关系，并以网格的形式表示。如果网格划分中存在明确的比例关系，可标注说明。
- 空间概念：标示由部件要素组织所限定的空间。

EVALUATION CRITERIA
- Compliance with program objectives and requirements.
- Use of entire field in the design.
- Demonstration of an understanding of datum.
- Quality of idea and composition

DRAWING DOCUMENTATION

Draw a 1:1 scale plan and three analytical diagrams of your design on two A3 drawing sheets. These diagrams are to be freehand drawings at half scale (1:2). Address the following issues with plan diagrams:
- Position of the datum object wall on the field
- Grid subdivision: Identify the major and minor subdivisions of the field and represent them as a grid of lines. Identify proportional relationships if they exist.
- Space concept: Identify the space/s that are defined by the organization of the elements.

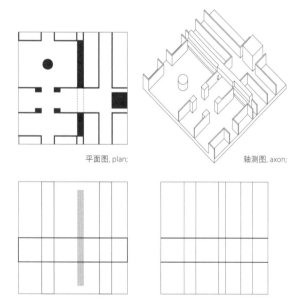

平面图, plan;　　　　　　　　　　　　　　　轴测图, axon;

空间关系, spatial relationship;　　　　　网格分割, grid subdivision;

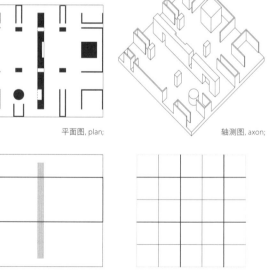

平面图, plan;　　　　　　　　　　　　　　　轴测图, axon;

空间关系, spatial relationship;　　　　　网格分割, grid subdivision;

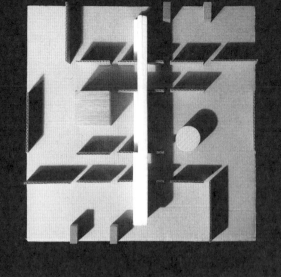
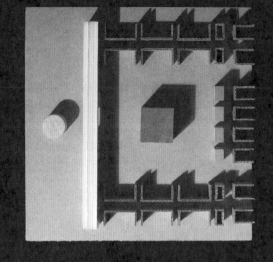

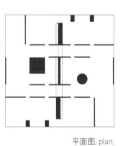

平面图, plan;

轴测图, axon;

空间关系, spatial relationship;

网格分割, grid subdivision;

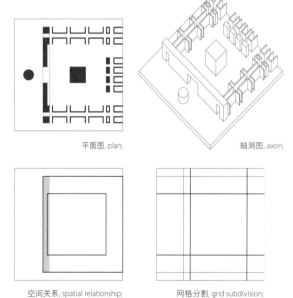

平面图, plan;

轴测图, axon;

空间关系, spatial relationship;

网格分割, grid subdivision;

125

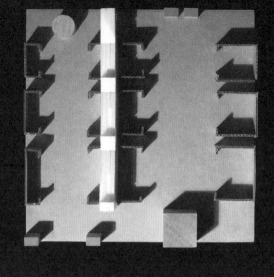

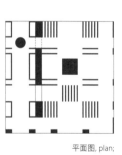

平面图, plan;

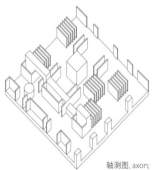

轴测图, axon;

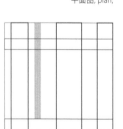

空间关系, spatial relationship;

网格分割, grid subdivision;

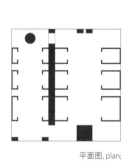

平面图, plan;

轴测图, axon;

空间关系, spatial relationship;

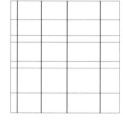

网格分割, grid subdivision;

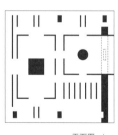
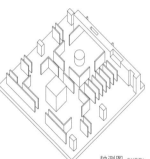

平面图, plan;

轴测图, axon;

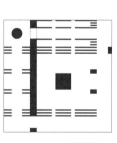
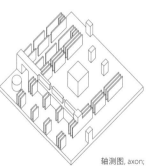

平面图, plan;

轴测图, axon;

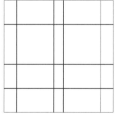

空间关系, spatial relationship;

网格分割, grid subdivision;

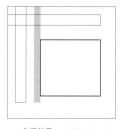
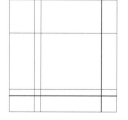

空间关系, spatial relationship;

网格分割, grid subdivision;

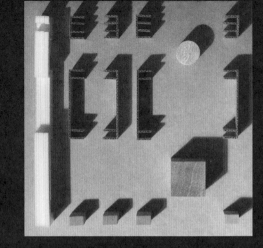

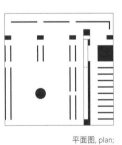

平面图, plan;

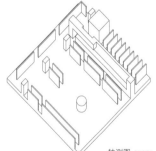

轴测图, axon;

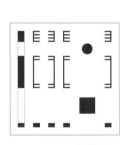

平面图, plan;

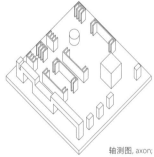

轴测图, axon;

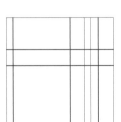

空间关系, spatial relationship;

网格分割, grid subdivision;

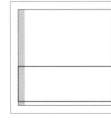

空间关系, spatial relationship;

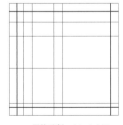

网格分割, grid subdivision;

128

LAM Cheung Ha

WU, Yu Shang Sunny

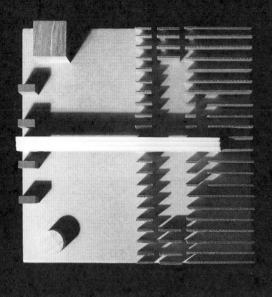

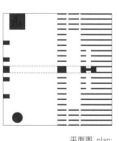

平面图, plan;

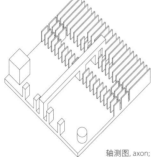

轴测图, axon;

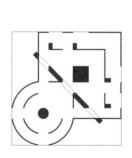

平面图, plan;

轴测图, axon;

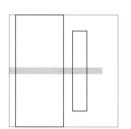

空间关系, spatial relationship;

网格分割, grid subdivision;

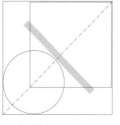

空间关系, spatial relationship;

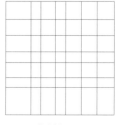

网格分割, grid subdivision;

小结 reflection

　　基准墙是一个分两步走的练习。第一步是设计一个尺寸给定（216mm x 36mm x 15mm）的墙体要素。通过一系列的挖去操作来创造洞口与壁龛，将这个统一且整体的墙体要素变为一个雕塑化的、相互关联的形式。练习对于该墙体的形式特征不做特别的限制，它可以是对称的或非对称的，一侧是平整的而另一侧是被"侵蚀"的，一端是实体的而另一端可以是体量感渐弱，或是其他任何可能。唯一的要求是控制其中挖去操作（创造洞口与壁龛）的数量，以及对墙体整体尺寸（长宽高）的保持。

　　在第二步中，将墙体布置在一块252mm x 252mm的底板上，墙体的摆放及其朝向可能会受到方形的几何属性的影响。设计中，需要将一组预先设计的给定部件要素布置在方形场地中。通过使用这些部件要素与墙体物件来创造一个整体的构成，其中墙体物件不再是独立而无关的单一要素，相反，它是场地构成的结果。将薄的片状墙体要素与之对位布置，继而生成一个网格系统，并以此来组织一组具有不同几何形式的物件实体，包括立方体、圆柱体与长方体。部件要素的形式、位置不仅与墙体物件相互关联，同时还成

Datum Wall is a two-part exercise. The first stage is to design a wall-object of a set dimension (216mm x 36mm x 15mm). This uniform and monolithic wall-object is transformed into a sculptural and articulated form through a subtractive process of creating holes and notches. The specific character of the wall has no specific formal constraint; it may be symmetrical or asymmetrical, flat on one side and eroded on the other, solid on one end and gradually dematerialized on the other, or one of many other possibilities. The only requirement is that a specific number of erosions (holes and notches) can be made and that the overall dimension of the wall (*l w h*) must be maintained.

In the second stage, the wall is placed on a square base (252mm). The exact position and orientation of the wall may be influenced by the geometrical properties of the square, however a *kit-of-parts* of pre-defined elements will also be deployed on the square field. These elements should work together with the wall-object to create an overall composition such that the articulations of the wall-object appear as a consequence of the field composition rather than independent and unrelated. Alignment of the thin, planar wall elements will go far in creating a grid that organizes the special geometric solids of the kit (cube, cylinder and small prisms) and relates their shape/position not only to the linear wall-object but also to the

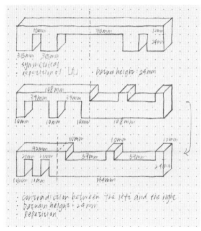

方案草图, sketch record (CHAU, Ka Yee Alice); 制作研究模型, work on study models;

为方形场地的一部分。

　　练习中所面临的挑战与设计问题的设定有关：创造一个强调形式"文脉主义"的构成。此外，它同时倡导一种"空间限定的模式"，在其中各部分空间均得到清楚的限定，并与墙体物件相互参照。同前面的各练习一样，在此需要重点考虑方形场地的划分问题。连同物件要素及各部件要素，对于场地进行层叠的、分层的阅读，可进一步加强设计图示的复杂性。基准墙练习中所渗透的组织策略（线性基准与场地基准）适用于任何尺度的设计，其中的共通性可以使之与城市设计问题关联。

overall field of the square.

In addition to the challenge of creating a composition that emphasizes a formal *contextualism*, the exercise also calls for "a pattern of space definition" with spaces that are well defined and reference the wall-object. In this problem, the subdivision of the square field continues as an important consideration and the potential for overlapping and layered *field* readings that both relate to the wall-object and the object elements of the kit-of-parts, contributes to the increasing complexity of the schemes. Resemblance of the datum wall projects to urban scale design is not unexpected as the same organizational strategies associated with linear and field datums can be employed at any scale.

3.4.1

网格框架 grid frame

任务书 handout

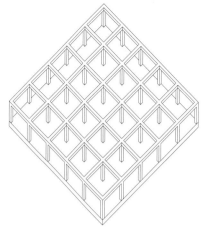

网格框架系统, grid frame system;

在一个由均等方形模数构成的网格结构中，使用一组给定的部件要素，限定一系列空间以创造空间的序列。

训练要求

在一块252mm x 252mm的场地上，布置有一个5格 x 5格的均跨网格框架，框架高36mm，由边长为5mm的方截面梁柱组成。框架覆盖于整个场地之上，柱表面相互平行，框架与底板的边角对齐。一组15个的部件要素需布置于网格框架中。要素可以与框架柱的表面相贴，但是两者不可相互穿插。除了这些要素外，还可考虑使用两种厚度不同的分隔墙体：厚墙厚度为5mm，薄墙厚度为2mm，两者高度均为24mm，长度不限。在框架的上部，可考虑添加矩形的屋面板片，屋面板同样分两种厚度。厚屋面板厚度为5mm，其顶面应与框架的顶面相平，即顶面距离模型底板36mm。薄屋面板厚度为2mm，它可以同厚屋顶板一样，顶面与框架的顶面相平布置，或是选择底面与梁的底面平齐，即板底离模型底板31mm。设计中应控制屋面板的数量：其覆盖面积不得超过模型底板总面积

Within a frame structure of equal square modules create a spatial sequence that incorporates a set of object elements and defines a series of spaces.

REQUIREMENTS

An equal bay 5 x 5 grid frame composed of square column and beam elements of 5mm x 5mm and a height of 36mm on a base 252mm x 252mm. The frame covers the entire base with column surfaces parallel and flush to the edges of the base. A kit of fifteen object-elements must be positioned within the grid frame. Elements may adjoin the surfaces of the frame but not intersect it. In addition to these elements, there are two types of wall partitions that may be added: thick walls and thin walls. The thick wall is 5mm x 24mm and variable length. The thin wall is 2mm x 24mm and also of variable length. Within the upper portion of the frame, rectangular roof planes may be introduced. Thick roof planes are 5mm thick and are positioned so that the upper surface is at the same height as the frame (36mm). Thin roof planes are 2mm thick and may be positioned with either the top surface aligned with the height of the frame (36mm) or the bottom surface with the lower edge of the beams at a height of 31mm. The only limitation on the amount of roof planes that may be used is that the covered area

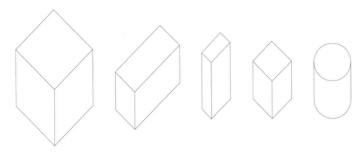

1件, 1 piece   36mm x 36mm x36mm
6件, 6 pieces   18mm x 24mm x42mm
4件, 4 pieces   9mm x 18mm x 36mm
3件, 3 pieces   18mm x 18mm x24mm
1件, 1 piece   24mmØ x 24mm

部件要素, object elements;

的30%。分隔墙与屋面板都应保持为矩形，其上不得开洞口或缺口。

由于网格框架本身已限定了一个5格 x 5格的矩阵空间模数，设计中可以通过以下方式的组合来进一步加强空间限定：

- 部件要素的摆放
- 水平屋面板的使用
- 非结构性垂直板片的使用

在此，空间序列并不要求有明确的开端和结束，也不要求塑造一条严格的线性路径。空间的序列可能是在创造一个模式，其中有一组具有层级关系的空间，空间之间的关系亦具有层级性。可以从底板的任意边上的任意点进入到这一序列及整个设计中。

评价标准

- 满足训练目标与要求
- 高效、经济的使用空间限定要素
- 空间组织概念的品质，部件要素与空间之于设计策略（构思）之间的关系
- 空间序列的趣味性与明确性

should be no more that 30% of the total area of the base. Partitions and roof planes are to be rectangular and solid, with no holes or notches.

While the grid frame itself defines a 5 x 5 matrix of space modules, spatial definition may be strengthened through any combination of the following:

- Placement of the object elements
- Use of the horizontal roof planes
- Use of the non-structural vertical planes

The spatial sequence is not required to have a beginning and an end, nor does it need to be configured as a strictly linear path. A sequence of spaces may form a pattern that has a hierarchy of spaces and relationships between the spaces. Access to the sequence and the overall design is from any point along one of the edges of the base.

EVALUATION CRITERIA

- Compliance with program objectives and requirements
- Effective and economic use of space defining elements
- Quality of the organizational concept and the relationship of the various elements and spaces to the controlling idea (parti)
- Interest and clarity of the spatial sequence

- 空间层次概念的发展
- 在一个变形的、有层级关系的网格系统中对于网格框架的整合
- 对于设计过程的记录：展现不同设计方案的生成以及设计概念的变化与发展
- 最终模型成果的手工品质

透明性研究

通过对网格框架设计方案进行二维的空间分析，以讨论其中空间的层叠与渗透。

在一块252mm x 252mm的白色纸板上进行拼贴研究。以方案的平面图为基础，使用白色、黑色、浅灰与深灰的纸片来表达其中不同的空间部分，以强调空间阅读。其中，黑色表示空间层叠关系最复杂的部分，深灰色次之，并以此类推，白色则表示实体要素（如部件要素、框架柱）以及稠密、近乎于实体的部分。在切割纸片时应尽可能准确，使其尺寸、位置与方案平面相符。然后在拼贴中，将这些纸片彼此相接排布，以创造出不同空间彼此层叠的效果。通过这个二维的拼贴研究，可以表现出设计空间之间的渗透，

- Development of the concept of layering
- Integration of the grid frame in a transformed and hierarchical grid system
- Documentation of the design process; evidence of the generation of alternative solutions as well as the transformation and development of a design concept
- Quality of craftsmanship in the presentation of the final project.

TRANSPARENCY STUDY

Create a two-dimensional spatial analysis of the grid frame design emphasizing the overlapping and interpenetration of space.

252mm x 252mm paper collage study on white cardboard. Use white, black and two tones of gray paper to "fill" the spaces in the plan of the grid frame scheme emphasizing the primary spatial readings. Black represents the most prominent overlapped space, dark gray the next most, and so forth. White represents solids between spaces (elements and columns) as well as areas that are dense and "almost" solid. Cut the papers precisely to match the dimensions and areas of the space in the plan. Place and attach the cut rectangular shapes next to each other in the collage to create the effect of two or more spaces overlapping. This is referred to as spatial inter-

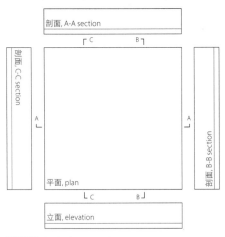

图纸排布, drawings' arrangment;

进而指向对空间透明性的识别。

penetration and leads to the recognition of spatial transparency.

绘图记录

　　在一张竖向放置的A2白色绘图纸上，排布一组徒手绘制的设计图纸与分析图解。其中，正投形图包括1幅平面图、1幅主要的立面图以及3幅剖面图，均以1:1的比例绘制。绘图时注意这几幅图纸之间的关系及方向。然后以1:2的比例绘制3张分析图解，以表达设计中涉及的相关问题，如空间层级关系、图底关系、网格划分、形式组织策略、空间序列等。

DRAWING DOCUMENTATION

　　Plan, elevation, section, and diagrammatic freehand drawings of the design on one A2 white paper (vertical). Multi-view drawings: 1 cut plan, 1 primary elevation and 3 sections are to be freehand drawings at full scale (1:1). Notice their relationship and orientation. Also draw three diagrams to indicated issues in the design, such as space hierarchy, figure/ground relationship, grid subdivision, formal organization concept, space sequence, etc. at half scale (1:2).

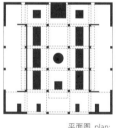

平面图, plan;

立面图, elevation;

轴测图, axon;

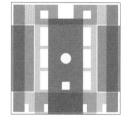

透明性研究, transparency study;

组织原则, ordering concept;

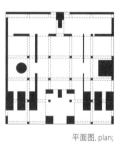

平面图, plan;

立面图, elevation;

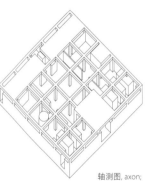

轴测图, axon;

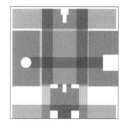

透明性研究, transparency study;

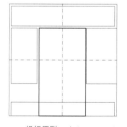

组织原则, ordering concept;

平面图, plan;

立面图, elevation;

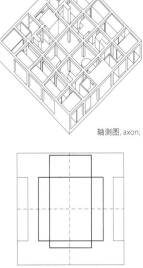

透明性研究, transparency study;

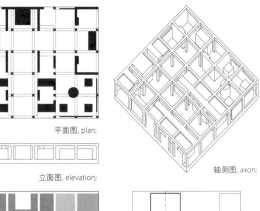

轴测图, axon;

组织原则, ordering concept;

平面图, plan;

立面图, elevation;

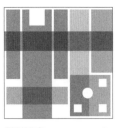

透明性研究, transparency study;

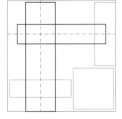

组织原则, ordering concept;

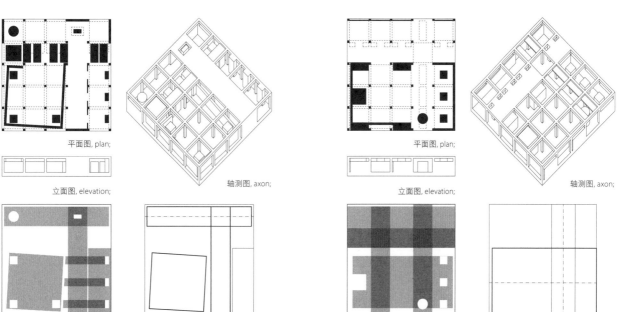

平面图, plan;

立面图, elevation;

轴测图, axon;

透明性研究, transparency study;

组织原则, ordering concept;

平面图, plan;

立面图, elevation;

轴测图, axon;

透明性研究, transparency study;

组织原则, ordering concept;

David STEIN

Tim BECHTOL

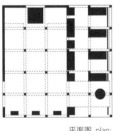

平面图, plan;

立面图, elevation;

透明性研究, transparency study;

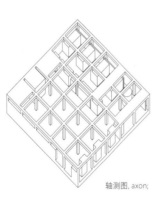

轴测图, axon;

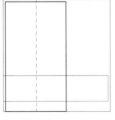

组织原则, ordering concept;

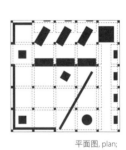

平面图, plan;

立面图, elevation;

透明性研究, transparency study;

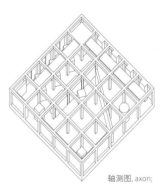

轴测图, axon;

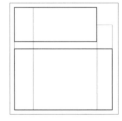

组织原则, ordering concept;

140

小结 reflection

练习的起点是一个给定的网格框架，设计需要与之进行互动。框架整体为方形，对边方向均分为5格 x 5格的方形网格单元。网格框架中的空间限定主要依赖于结构柱间置入的分隔墙。当多个模数单元集合成为群组时，便可以获得相对大的空间。结构与空间限定之间的这种"契位"限定关系便是模块式的、填充的空间设计。由此生成的空间可能是腔体状的（分隔墙垂直满布于楼板与顶棚之间），或是开放平面（分隔墙的高度有限，网格单元之间存在视觉联系）。

在这个练习中，所有的分隔墙高度均为24mm，其顶端距离框架上表面12mm。这样，也就暗示了开放平面的空间特点。部件要素包括一个大的立方体，一个圆柱体以及多个尺寸不同的长方体。另外，有两种不同厚度的板片要素（2mm与5mm），它们可以作为分隔墙体或屋顶板片使用。通过布置这些部件及板片要素，可以创造出一个空间与流线路径均得到清楚限定的设计。

空间在尺寸与配置关系上的变化，将会与网格框架本身的规则性产生对比。要素的布置可以选择与网格相对位或异位，由此往往会在两个系统间产生一定

A grid frame is provided in this exercise as a starting point, as something to react against. The overall frame is square with a 5 x 5 grid of square bays. Space definition within a grid frame structure is typically made by inserting partitions between structural columns. Larger spaces are possible by grouping modules together. This type of *close fit* between structure and space definition is referred to as modular or *infill* spatial design. The character of the spaces formed may be *cellular*, with partition walls extending from floor to ceiling, or *open plan*, in which partition walls are keep low so that there is visual connection through the modules of the frame.

In this exercise all partitions are at a height of 24mm, or 12mm below the top surface of the frame. Therefore, an open plan space character is implied. Elements of the kit of parts include a large cube, a cylinder and various rectangular prisms. In addition, planar elements of two thicknesses (2mm and 5mm) may be created for use as wall partitions or roof planes. These objects and planes are employed to create a pattern of well-defined spaces and circulation paths.

The variation in the size and configuration of the spaces will contrast with the regularity of the grid frame. Alignments may be on-grid or off-grid and there will generally be some overlap between the two systems. Larger field readings will be implied. Edges of a field may be articulated in several

灰度色卡, monochrome color chart;

的交叠。这也暗示了区域阅读的进一步扩展。区域边界的确定也有不同的实现方式：通过屋顶要素、实体物件要素的摆放或者是网格框架本身。

区域间的交叠也会加强或限定彼此交叠的空间带。空间带的清楚限定以及对于它们的解读，可以引入"空间层次"的概念。在此，可以使用二维的图形再现——空间的透明性拼贴，来研究空间的层叠模式。通过不同灰度的卡纸（由浅灰到黑色，如灰度色卡所示）来创造一个图式，以最深的颜色（黑色）表示其中空间层叠最密集的区域，而以递减的灰度表示空间带限定渐弱、层叠关系愈简的区域。该练习的目的是在设计的发展过程中，甄别与强化主要空间区域的阅读。它应当作为一个有效的设计工具，参与到丰富且清晰可读的空间设计创造中。二维图形再现将有助于减弱具有重复性结构模数的网格框架的影响，并促生一个开放的、较少限制的空间配置关系。

屋顶要素的参与，可以为空间限定以及网格线参照带来了另一维度上的变化。作为屋顶元素的板片支承于梁的侧边、填充于框架之中。应保证，矩形板片至少有一对边与框架梁相连接。填充厚度为5mm的屋

ways; by roof elements, positioning of solid object elements, or by the grid frame itself.

The overlapping of fields will also reinforce or define overlapping spatial zones. The precise definition of spatial zones and their possible *interpenetration* introduces the concept of spatial layering. The pattern of space overlap can be studied in a two-dimensional graphic representation, the *spatial transparency collage*. Using tones of gray paper ranging from light gray to black (e.g., monochrome color paper), a pattern can be created in which the darkest shade (black) represents the most intensified spatial overlap, while lesser defined spatial zones and overlaps are given lighter tones of gray in descending order. This exercise focuses on identifying and reinforcing the readings of key spatial areas in the developing design. It is best incorporated as a design tool in creating a complex yet legible spatial design. The two-dimensional graphic representation helps to diminish the influence of the grid frame with its repetitive structural module and promote a more open and less restricted spatial configuration.

Roof elements add another dimension to the spatial definition and grid line reference. Planes employed as roof elements are considered infill, being supported only on the sides of the beams. At least two opposite edges of a rectangular plane must attach to the beams of the frame. The thick (5mm)

顶板片时，应使之与等厚的框架梁平齐。而厚度为2mm的屋顶板片与框架梁之间，可以有多种位置关系：两者的上表面平齐或下表面平齐，或是将板片置于梁高的中间位置。使用5mm厚屋顶板片进行填充，或是将2mm厚屋顶板片与框架梁底平齐布置时，其底面便会表现为平滑的连续表面，从而消解框架的模块性。虽然框架柱依然存在，但框架梁则趋于中性。训练中，屋顶的覆盖面积限制为整体顶面的30%，以此保证开放空间与覆顶空间之间的平衡，否则屋顶板片便丧失了与设计练习之间的关联。

planes can fill the space of the frame between the beams that are also 5mm in depth. Thin planes (2mm) may be placed in a variety of positions: top surface aligned with top edge of beam, bottom surface aligned with the bottom edge of beam or in the middle. Using 5mm planes or aligning the 2mm planes with the bottom edge of the beams will tend to produce a smooth, continuous ceiling on the underside that will tend to minimize the modularization of the frame (the beams are neutralized although the columns of the frame remain). A limit on the amount of roof plane coverage is placed at 30%. There should be a balance between open versus covered spaces otherwise the roof plane loses its relevance to the exercise.

3.4.2
网格构成 grid composition

任务书 handout

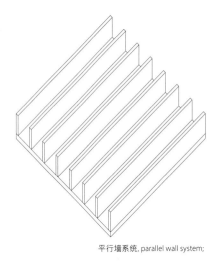

平行墙系统, parallel wall system;

在一个方形场地上创造一个设计，并从中来探索网格组织的潜力。场地上沿单一方向布置有一组平行墙结构，它为整个构成确定了主要的朝向。练习的挑战在于，使用给定的墙体与屋顶板片来发展一个横纵正交的网格。该网格构成应该具有层级性，并包含一系列清晰限定的、多样的空间。

训练要求

场地252mm × 252mm，其上布置高36mm，长度不少于12mm的矩形垂直板片。垂直板片用以支承水平板片，两者顶面平齐。水平板片最小宽度为12mm，总面积不得超过场地底板面积的30%。需保持上述板片为矩形实体，其上不可开洞口或缺口，材料为厚5mm的瓦楞纸板。垂直板片需布置在8条平行线上，且其中的两条紧贴底板两对边边缘。每两片平行板片之间的净距离不得少于18mm。

在与垂直板片（厚5mm）相正交的方向上布置高24mm的薄的垂直板片，以形成一个网格系统。对于这些薄片的使用，没有数量或长度的限制。这些薄片可以用来支承薄的水平板片，需保证两者顶面平齐。所

On a square field create a design that explores the potential of grid organization. The composition will have a dominant field orientation established by a parallel wall structure in one direction. The challenge is to develop a cross-grain gridding in the perpendicular direction using walls and roof planes as proscribed in the exercise. The grid composition should be hierarchical and contain a variety of well-defined spaces.

REQUIREMENTS

One 252mm × 252mm base. Rectangular vertical planes 36mm high with a minimum length of 12mm. These vertical planes may support horizontal planes on their upper edges with a minimum width of 12mm and whose total area is less than 30% of the area of the base. All planes are to be solid with no holes or notches and will be made of 5mm corrugated cardboard. The vertical planes are to be positioned in 8 parallel lines, two of these lines must be coincident with the edges of the base. The distances between any two parallel vertical planes may vary but can be no closer than 18mm.

In the direction perpendicular to the thick (5mm) vertical planes, a grid of thin vertical planes at a height of 24mm can be introduced. There is no restriction on the number or length of these thin vertical planes. These planes may support thin horizontal planes on their upper edges. Both

薄片同样为矩形实体，不可开挖洞口或缺口，材料为厚2mm的灰卡纸。

练习中所创造的空间在尺寸、比例以及围合程度上应有所差别。构成中没有水平板片覆盖的部分可以被感知为虚空，将这些空间视为"室外房间"。

评价标准

- 满足训练目标与要求
- 清晰、明确的空间限定
- 构成的品质：场地划分，网格之组织结构，形式组织构思
- 使用秩序组织原则来强化构成
- 最终模型成果的手工品质

绘图记录

在两张A3绘图纸上，使用徒手线条分别绘制1幅1:1的平面图及3幅1:2的平面分析图解，作为设计记录。分析图解可围绕下述问题展开

- 实体与空间：对图底关系的解读，设计中有水平板片覆顶的区域可以认为是实体部分；

vertical and horizontal thin planes are to be solid with no holes or notches and will be made of 2mm gray cardboard.

Spaces that are created may vary in size, proportion and degree of enclosure. They are to be perceived as voids in the composition, that is, without the horizontal plane closure. Think of the spaces as "outdoor rooms".

EVALUATION CRITERIA

- Compliance with program objectives and requirements.
- Clear and unambiguous spatial definition.
- Quality of the composition: subdivision of the field, grid as organizing structure, strength of parti.
- Use of ordering principles to strengthen the composition.
- Quality of craftsmanship in the presentation of the final project.

DRAWING DOCUMENTATION

Draw a 1:1 scale plan and three analytical diagrams of your design on two A3 drawing sheets. These diagrams are to be freehand drawings at half scale (1:2). Address the following issues with plan diagrams

- Solid-void: A figure/ground interpretation. The area with horizontal planes in the design may be conceived of as solid areas.

－ 网格分割：区分场地划分的主次层级关系，并以网格的形式表示，如果网格划分中存在明确的比例关系，可标注说明；

－ 组织原则：图解表示构成操作的组织原则，如基准、轴线、组群、几何关系等。

- Grid subdivision: Identify the major and minor subdivisions of the field and represent them as a grid of lines. Identify proportional relationships if they exist.

- Organizational concept: Illustrate the principal ordering concept(s) that structure your design (for ex. - datum, axis, hierarchy, geometry, etc.)

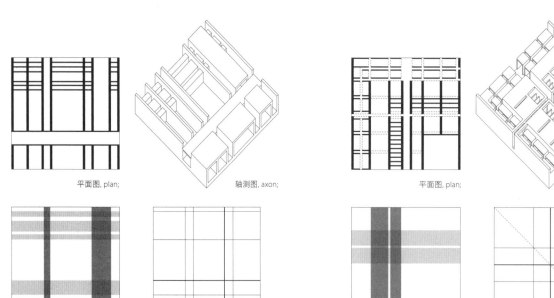

平面图, plan;　　　　　　　　　　　　　　　　　　　　軸測图, axon;　　　　　　　　　平面图, plan;　　　　　　　　　　　　　　　　　　　軸測图, axon;

空间关系, spatial relationship;　　　　　　　网格分割, grid subdivision;　　　　　　空间关系, spatial relationship;　　　　　　网格分割, grid subdivision;

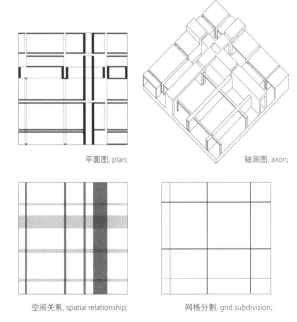

平面图, plan;　　　　　　　　　　轴测图, axon;

空间关系, spatial relationship;　　　　网格分割, grid subdivision;

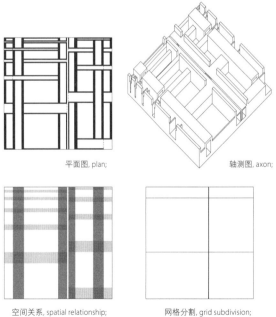

平面图, plan;　　　　　　　　　　轴测图, axon;

空间关系, spatial relationship;　　　　网格分割, grid subdivision;

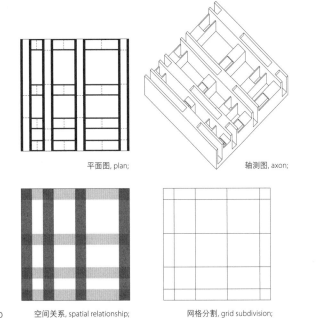

平面图, plan;　　　　　　　　　　　轴测图, axon;

空间关系, spatial relationship;　　　　网格分割, grid subdivision;

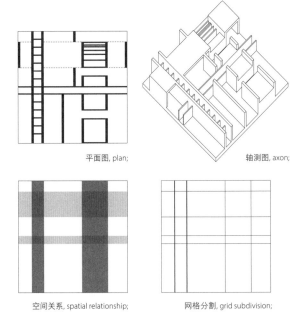

平面图, plan;　　　　　　　　　　　轴测图, axon;

空间关系, spatial relationship;　　　　网格分割, grid subdivision;

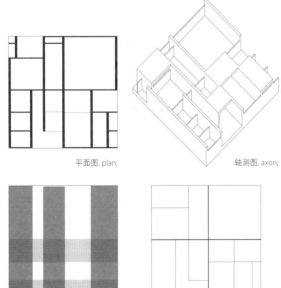

平面图, plan;

轴测图, axon;

空间关系, spatial relationship;

网格分割, grid subdivision;

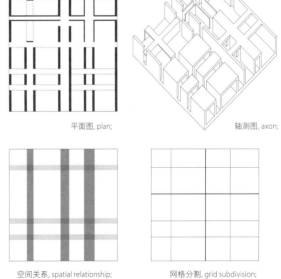

平面图, plan;

轴测图, axon;

空间关系, spatial relationship;

网格分割, grid subdivision;

3.4.3
平行墙 – 物件 parallel wall - objects

任务书 handout

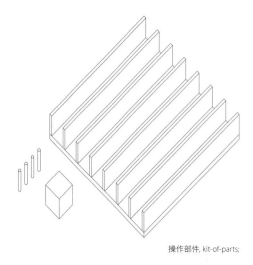

操作部件, kit-of-parts;

　　在方形场地上，使用垂直平行板片、水平板片以及一组部件要素进行空间的限定与组织。空间之间应存在层级关系，并在设计中表达空间层次的概念。

训练要求

　　练习中，使用一块252mm x 252mm的模型底板；高36mm的矩形垂直板片，宽度不得小于12mm；矩形水平板片最小边宽为12mm，其覆盖范围不可超过底板面积的30%。所有的板片上不得开洞口或缺口。水平板片需对边支承在垂直板片上，两者的顶面平齐，距离模型底板表面36mm。垂直板片需布置在与底板单一方向平行的8条平行线上，且其中两条紧贴底板两对边边缘。设计中，可以调整平行板片之间的距离，但不得少于18mm。

　　除了垂直板片和水平板片外，还有5个部件要素需要被有效地组织到设计中。这些要素包括一块边长为36mm的立方体，以及4根高36mm、直径为5mm的细圆柱。在组织要素的空间关系时，可从以下两个角度考虑其作用：

　　－　将物件用于空间占据（可独立布置，甚至成为

On a square field create a design using parallel vertical planes, horizontal planes and a set of object elements, which defines a hierarchical group of spaces and incorporates the concept of spatial layering.

REQUIREMENTS

One 252mm x 252mm base. Rectangular vertical planes 36mm high with a minimum dimension in width of 12mm. Rectangular horizontal planes with a minimum width of 12mm and whose total area is less than 30% of the area of the base. All planes are to be solid with no holes or notches. Horizontal planes must be supported along two opposite edges at a height of 36mm. Vertical planes are positioned in 8 parallel lines; two of these lines must be coincident with the edges of the base. The distance between any two parallel vertical planes may vary but can be no closer than 18mm.

In addition to the vertical and horizontal planes, there are five object elements that are to be strategically introduced into the design: a 36mm high cube and four 36mm high round columns that are 5mm in diameter. Two possible spatial relationships for these elements in the design might be possible.

－　Objects are used as space occupiers (freestanding and perhaps

中心焦点）

－ 将物件用于空间限定（角落或边缘严格对位）

构成中应善用场地的网格布局与"实体－空间连续体"来清楚限定、组织空间。我们可以清楚感知这些空间为虚空，比如说不使用水平板片对其覆盖包裹。水平板片会带来密实的感觉。设计应将"室外房间"作为主要的空间考虑。

所有的片状部件要素（包括水平板片与垂直板片）均使用厚5mm的瓦楞纸板，底板及其他部件要素的材料为中密度纤维板。

评价标准

－ 满足训练目标及要求
－ 清晰明确的空间限定
－ 对于空间层次的认识及应用
－ 使用秩序组织原则来强化构成
－ 最终模型成果的手工品质

绘图记录

在两张A3绘图纸上，使用徒手线条分别绘制1幅1:1

focal).

- Objects are used as space definers (corner/edge alignments critical).

The group of spaces required in the composition should be well defined yet enmeshed within the gridding and solid-void continuum of the field. These spaces should be readily perceived as voids, i.e., do not use horizontal plane closure. Use the horizontal planes to introduce density and solidity. Think of the dominant spaces as "outdoor rooms".

All planar elements, both vertical and horizontal planes, can be made from corrugated cardboard approximately 5mm in thickness. Base and object elements are made of MDF.

EVALUATION CRITERIA

- Compliance with program objectives and requirements.
- Clear and unambiguous spatial definition.
- Introduction and use of layering.
- Use of ordering principles to strengthen the composition.
- Quality of craftsmanship in the presentation of the final project.

DRAWING DOCUMENTATION

Draw a 1:1 scale plan and three analytical diagrams of your design on

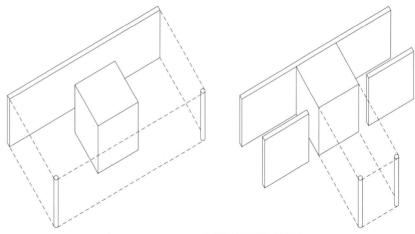

物件用于空间占据与空间限定, space occupier and space definer;

的平面图以及3幅1:2的平面分析图解，作为设计记录。分析图解可围绕下述问题展开。

– 实体与空间：对图底关系的解读，设计中有水平板片覆顶的区域可以认为是实体部分；

– 网格分割：区分场地划分的主次层级关系，并以网格的形式表示，如果网格划分中存在明确的比例关系，可标注说明；

– 组织原则：图解表示构成操作的组织原则，如基准、轴线、组群、几何关系等。

two A3 drawing sheets. These diagrams are to be freehand drawings at half scale (1:2). Address the following issues with plan diagrams

- Solid-void: A figure/ground interpretation. The area with horizontal planes in the design may be conceived of as solid areas.

- Grid subdivision: Identify the major and minor subdivisions of the field and represent them as a grid of lines. Identify proportional relationships if they exist.

- Organizational concept: Illustrate the principal ordering concept(s) that structure your design (for ex. - datum, axis, hierarchy, geometry, etc.)

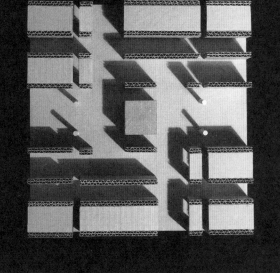

平面图, plan;

轴测图, axon;

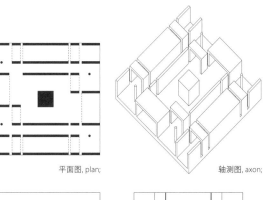

平面图, plan;

轴测图, axon;

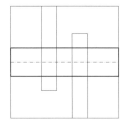

空间关系, spatial relationship;

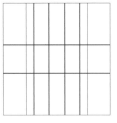

网格分割, grid subdivision;

空间关系, spatial relationship;

网格分割, grid subdivision;

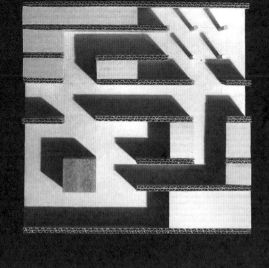

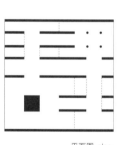

平面图, plan;

轴测图, axon;

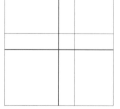

空间关系, spatial relationship;

网格分割, grid subdivision;

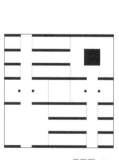

平面图, plan;

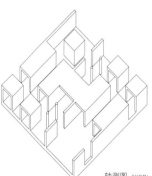

轴测图, axon;

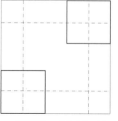

空间关系, spatial relationship;

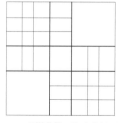

网格分割, grid subdivision;

157

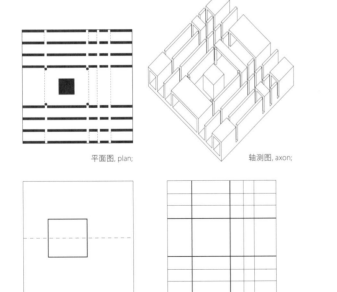

平面图, plan;

轴测图, axon;

空间关系, spatial relationship;

网格分割, grid subdivision;

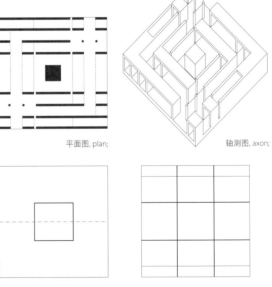

平面图, plan;

轴测图, axon;

空间关系, spatial relationship;

网格分割, grid subdivision;

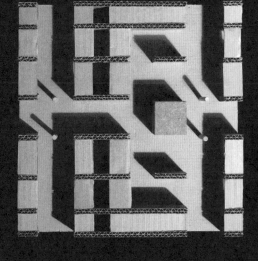

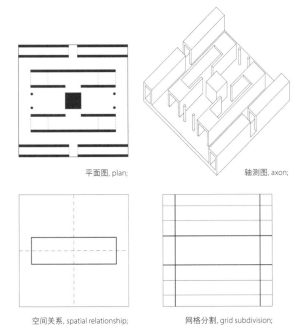

平面图, plan;

轴测图, axon;

空间关系, spatial relationship;

网格分割, grid subdivision;

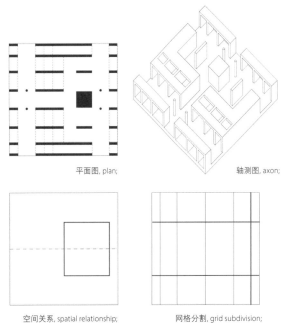

平面图, plan;

轴测图, axon;

空间关系, spatial relationship;

网格分割, grid subdivision;

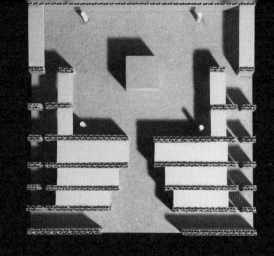

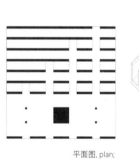

平面图, plan;

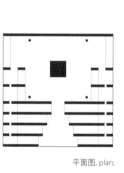

轴测图, axon;

平面图, plan;

轴测图, axon;

空间关系, spatial relationship;

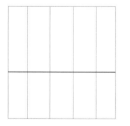

网格分割, grid subdivision;

空间关系, spatial relationship;

网格分割, grid subdivision;

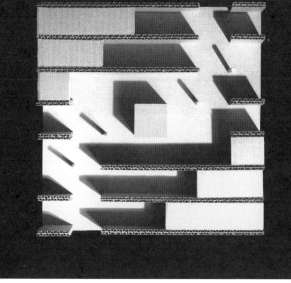

平面图, plan;

轴测图, axon;

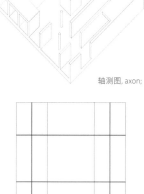

空间关系, spatial relationship;

网格分割, grid subdivision;

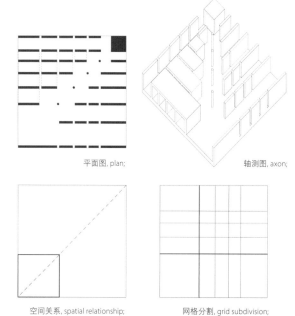

平面图, plan;

轴测图, axon;

空间关系, spatial relationship;

网格分割, grid subdivision;

小结 reflection

在这两个设计研究（3.4.2与3.4.3）中，结构被视作形式组织的工具，替代网格框架的是一组相互平行的墙体。这组墙体最初表现为中性的、等距布置的网格，与5格×5格的网格框架相似。面对这一给定的条件，学生可以通过改变墙体间的间距来创造一个具有层级关系的空间系统。在第一个练习"网格构成"中，引入了两种要素来加强空间限定：薄且低矮的板片（在此视为非结构性的分隔墙体），以及水平的屋顶板片（支承于结构性的平行墙之间）。屋顶板片呈现的是实体性，以此，开放、无覆顶的空间与覆顶、相对密闭的区域之间便产生了对比。事实上，这些覆顶的区域是密实的空间，在这个练习中应当视作实体存在。

平行墙之间成比例的间距布置可以创造出有趣的设计可能。密布墙体（最小净间距为18mm）且上覆屋顶要素，我们可以将它阅读为"厚墙"，就墙体系统本身创造层级关系，并区隔两侧的开放空间。将墙体成组组合，可以减少独立墙体板片的数量，并创造出更大的开放空间。将墙体紧密的成组布置，可以创造出设计中相对密实的区域，其角色相当于练习3.3.2"空间限定"中的"剖碎"。

These two design studies (3.4.2 and 3.4.3) introduce structure as a formal organizational device. In place of a grid frame structure is a series of parallel walls. The walls initially present themselves as a neutral, evenly spaced grid, much like the 5 x 5 grid frame. Reacting against this given condition the student is encouraged to vary the spacing between walls in order to develop a hierarchical system of spaces. In the first of the two exercises, Grid Composition, two other elements are introduced to enhance space definition: thin, low height vertical planes (conceived here as non-structural wall partitions) and horizontal roof planes (supported between the structural parallel wall planes). The roof planes create the appearance of solidity and hence help to establish contrast between the open, non-roofed spaces and the covered, more enclosed roofed areas. These covered areas are in fact, enclosed spaces but in this exercise should be envisioned as solids.

The proportional spacing of the parallel walls can lead to many interesting possibilities. Walls closely spaced (18mm minimum spacing) with roof elements read as *thick walls*, creating hierarchy within the system of walls and separation between open spaces. Combining the walls in pairs reduces the number of freestanding wall planes and allows for larger open spaces to be formed. Walls that are grouped close together create denser areas of the design and can assume the role of *poché* as in exercise 3.3.2 (space definition).

制作研究模型, work on study models; 绘制模型草图, make sketch of study model;

　　与厚的平行墙相对的是较薄、较短的板片，它们与平行墙体相互正交布置。将两种墙体配合在一起，可以创造出一个具有明显格纹的层级化网格系统。结构墙体上的开口暗示了穿行于空间之间的运动，同等重要的是，它在视觉上促进了两套网格系统间的相互渗透。屋顶要素间的间隔进一步强化了十字格纹的阅读。在建立区域中，屋顶要素同样有重要的作用：清楚限定一个由空间与实体共同组成的矩形区域。

　　在练习3.4.3中，需要对同一组平行墙进行变形：改变墙体间距、引入墙体间的间隔或是移除部分墙体片段。如练习3.4.2一样，覆于墙体之间的屋顶板片创造了实体与虚空之间的对比。练习中新加入的1个立方体与4根圆截面柱是非常重要的设计要素，可用以满足不同的设计需求。首先，作为独立要素，它们可以充当标志物或具有指引性的焦点。其次是空间性的作用，立方体的侧边或是占据矩形角点的4根立柱可以对空间进行限定。若将这些立柱相互对位布置，可以认为是一堵挖去了部分片段的实墙。不论设计构思如何，方形场地的几何属性是不可忽视的，设计应当在认识其基本属性的基础上，充分利用整个方形场地。

Working in contrast to the thick parallel walls are the thin shorter planes, positioned perpendicular to the parallel walls. Together, both sets of walls create a strongly hierarchical grid with a distinct grain. Openings in the structural walls suggest movement through and between spaces but are also important in facilitating the *interpenetration* of the two grids visually. Gaps in the roof elements further reinforce *cross grain* readings. The roof elements also play an important role in establishing *fields*; rectangular areas that are well defined and may consist of both space and solids.

In Exercise 3.4.3, the same set of thick parallel walls are again transformed by altering the spacing and introducing gaps, or sections of wall removed. As in Exercise 3.4.2, roof planes spanning between the walls create a contrast between solid and void. In this exercise, a cube and four columns are introduced and become important elements in the design. These object elements can perform several tasks. In the first instance they can remain freestanding objects acting as markers or directional foci. A second role is more spatial. The side of the cube or four columns in a rectangle or square pattern can act as space definers. Alternatively the columns can align and suggest "a wall with voids". In any configuration or *parti*, the geometry of the square field is unavoidable. The design should strive to engage the entire square field and acknowledge the fundamental properties of the square.

3.4.4
平行墙 parallel wall

任务书 handout

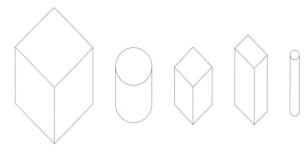

1件, 1 piece   36mm x 36mm x36mm
1件, 1 piece   24mmø x 24mm
3件, 3 pieces   18mm x 18mm x24mm
4件, 4 pieces   12mm x 18mm x 36mm
4件, 4 pieces   5mmø x 36mm

部件要素, object elements;

在一块平行墙的场地上，结合一组要素物件来进行空间序列设计。

训练要求

在252mm x 252mm的场地上，布置8片高36mm、长度不超过252mm的垂直平行墙，平行墙应始终保持与底板的两对边平行。位于尽端的两片平行墙应贴合底板边缘，且墙面应尽可能保持完整，无断开或洞口，可在两者中择其一开洞口作为场地入口及空间序列的起点。这8片平行墙是设计的主要结构构件，可用以支承水平"屋面"板片。可以考虑将这些平行墙截断以创造墙之间的间隙，或在其上开门洞。门洞或通道高度为24mm，宽度在6mm~72mm之间。平行墙之间的净间距不得小于18mm。

其他的空间限定要素包括：

－ 部件要素（如上图所示）。

－ 水平板片。覆顶部分面积不得超过底板总面积的30%。这些板片为矩形，短边尺寸不得小于12mm，其上不得开洞口或缺口。板片两对边支承在平行墙之上（不可悬挑），保证5mm厚的水平板片顶面与垂直

Within a field of parallel walls design a spatial sequence that incorporates a set of object elements.

REQUIREMENTS

Eight parallel walls 36mm in height and up to 252mm in length on a base 252mm x 252mm. Walls must be parallel with two edges of the base and two walls must be positioned along the edge and be mostly solid (no breaks or openings) with the exception that one wall must allow for entry and signal the beginning of the sequence. The parallel walls (all eight) are the primary structural elements and may support horizontal planes "roof" planes. The walls may also be interrupted (i.e. gaps in the wall) or may contain "portal" openings. These portals or doorways must be 24mm in height and between 6mm and 72mm in length. The minimum dimension of the space between the walls is 18mm.

Other space defining elements are:

- Object elements (see specifications below).
- Horizontal planes. Maximum total area less than 30% of the base. These are rectangular planes with no notches or holes and at least 12mm in the shortest dimension. They are supported on opposite edges by the primary structural planes (i.e. no cantilevers) and are either at a height of

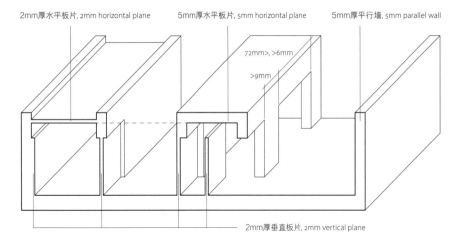

2mm厚水平板片, 2mm horizontal plane   5mm厚水平板片, 5mm horizontal plane   5mm厚平行墙, 5mm parallel wall

72mm>, >6mm

>9mm

2mm厚垂直板片, 2mm vertical plane

要素及操作规则, elements and operation limits;

板片顶面平齐，即其定位距离底板36mm，或使用2mm厚的水平板片，保持其底面与5mm厚水平板片的底面平齐，详见上图解。

   – 非结构性垂直板片。这些垂直板片高24mm、厚2mm。所有的垂直板片应保持为矩形，其上不得开洞口或缺口，短边尺寸不得少于12mm。板片的长边跨度不超过36mm。由于这些垂直板片属于非结构性构件，所以不可以使用它们来支承上述的水平屋面板。它们可独立布置、依附于平行墙或插入平行墙的洞口中（如上图解所示）。

   练习中可以通过以下方式的组合来限定空间：
   – 平行墙之间距变化（不小于18mm）
   – 部件要素的布置
   – 平行墙的挖去操作（通过移除墙体片段以创造间隙或通道）
   – 在主要的墙结构，即平行墙上开挖门洞。门洞高为24mm，宽度在6mm与72mm之间，相邻洞口的净间距不得小于9mm
   – 使用水平板片

36mm and flush with the top of the walls (5mm planes) or slightly below the top of the walls and aligning with the underside of the 5mm planes (2mm planes). See descriptive illustration.

   - Non-structural vertical planes. These vertical planes are 24mm in height and 2mm in thickness. They are also rectangular with no notches or holes and at least 12mm in the shortest dimension. The long dimension is limited to a span of 36mm. They are strictly non-structural and may not be used to support horizontal roof planes. They may be freestanding, attached to walls, or inserted into openings in the parallel walls. See descriptive illustration above.

Spatial definition may result through any combination of the following:
   - Variation of the distance between the walls (18mm minimum)
   - Placement of the object elements
   - Subtraction of the primary walls (removing segments to create gaps)
   - Creation of "portal" openings in the primary wall structure. The portals must have a height of 24mm, a width of between 6mm and 72mm, and a minimum distance between adjacent portals of 9mm
   - Use of the horizontal planes
   - Use of the non-structural vertical planes

－ 使用非结构性垂直板片

所有的部件要素均应打磨后，刷上白色石膏。

评价标准

－ 满足训练目标与要求

－ 高效、经济的使用空间限定要素

－ 空间组织概念的品质，部件要素与空间之于设计策略（构思）之间的关系

－ 空间序列的趣味性与明确性

－ 空间层次概念的发展

－ 对于网格组织的综合运用

－ 对于设计过程的记录：展现不同设计方案的生成以及设计概念的变化与发展

－ 最终模型成果的手工品质

透明性研究

通过对平行墙设计方案进行二维的空间分析，以讨论其中空间的层叠与渗透。

在一块252mm x 252mm的白色纸板上进行拼贴研究。以方案的平面图为基础，使用白色、黑色、浅灰

All elements to be sanded and painted flat white (gesso).

EVALUATION CRITERIA

- Compliance with program objectives and requirements

- Effective and economic use of the space defining elements

- Quality of the organizational concept and the relationship of the various elements and spaces to the controlling idea (parti).

- Interest and clarity of the spatial sequence.

- Development of the concept of layering.

- Integration and development of the concept of grid

- Documentation of the design process; evidence of the generation of alternative solutions as well as the transformation and development of a design concept.

- Quality of craftsmanship in the presentation of the final project.

TRANSPARENCY STUDY

Create a two-dimensional spatial analysis of the parallel wall design emphasizing the overlapping and interpenetration of space.

252mm x 252mm paper collage study on white cardboard. Use white, black and two tones of gray paper to "fill" the spaces in the plan of the

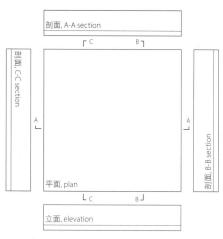

剖面, A-A section

剖面, C-C section

C

B

A

A

剖面, B-B section

平面, plan

立面, elevation

C

B

图纸排布, drawings' arrangment;

与深灰的纸片来表达其中不同的空间部分，以强调空间阅读。其中，黑色表示空间层叠关系最复杂的部分，深灰色次之，并以此类推，白色则表示空间之间的实体部分（如平行墙）以及稠密、近乎于实体的部分。在切割纸片时应尽可能准确，使其尺寸、位置与方案平面相符。然后在拼贴中，将这些纸片彼此相接排布，以创造出不同空间彼此层叠的效果。通过这个二维的拼贴研究，可以表现出设计中空间之间的渗透，进而指向对空间透明性的识别。

绘图记录

在一张竖向放置的A2白色绘图纸上，排布一组徒手绘制的设计图纸与分析图解。其中，正投形图包括1幅平面图、1幅入口侧立面图以及3幅剖面图，均以1:1的比例绘制。绘图时注意这几幅图纸之间的关系及方向。然后以1:2的比例绘制3张分析图解，以表达设计中涉及的相关问题，如空间层级关系、图底关系、网格划分、形式组织策略、空间序列等。

parallel wall scheme emphasizing the primary spatial readings. Black represents the most prominent overlapped space, dark gray the next most, and so forth. White represents solids between spaces (walls) as well as areas that are dense and "almost" solid. Cut the papers precisely to match the dimensions and areas of the space in the plan. Place and attach the cut rectangular shapes next to each other in the collage to create the effect of two or more spaces overlapping. This is referred to as spatial interpenetration and leads to the recognition of spatial transparency.

DRAWING DOCUMENTATION

Plan, elevation, section, diagram freehand drawings of the design on one A2 white paper (vertical). Multi-view drawings: 1 cut plan, 1 entrance side elevation and 3 sections are to be freehand drawings at full scale (1:1). Note their relationship and orientation. And also draw three diagrams to indicated issues in the design, such as space hierarchy, figure/ground relationship, grid subdivision, formal organization concept, space sequence, etc. at half scale (1:2).

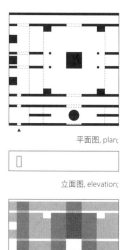

平面图, plan;

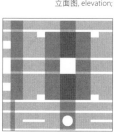

立面图, elevation;

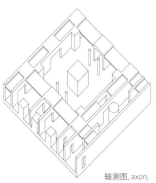

轴测图, axon;

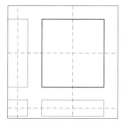

透明性研究, transparency study;

组织原则, ordering concept;

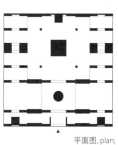

平面图, plan;

立面图, elevation;

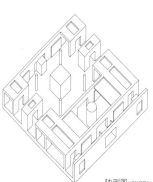

轴测图, axon;

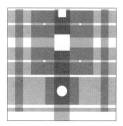

透明性研究, transparency study;

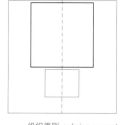

组织原则, ordering concept;

169

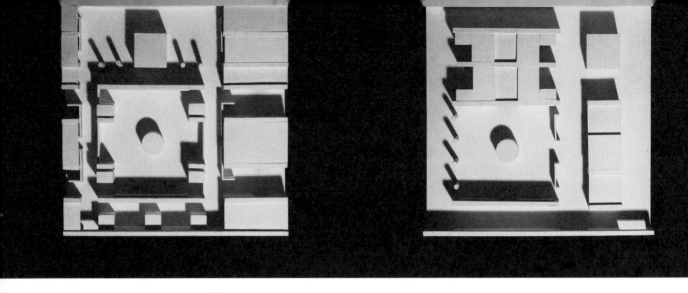

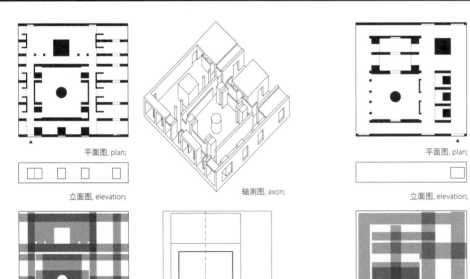

平面图, plan;

立面图, elevation;

轴测图, axon;

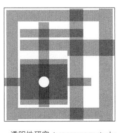

透明性研究, transparency study;

组织原则, ordering concept;

平面图, plan;

立面图, elevation;

轴测图, axon;

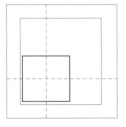

透明性研究, transparency study;

组织原则, ordering concept;

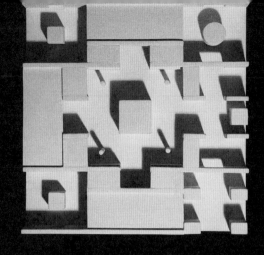

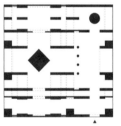

平面图, plan;

立面图, elevation;

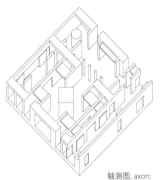

轴测图, axon;

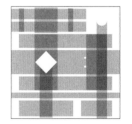

透明性研究, transparency study;

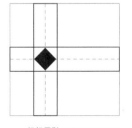

组织原则, ordering concept;

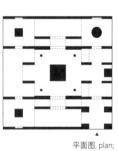

平面图, plan;

立面图, elevation;

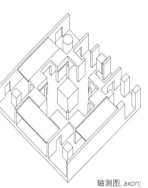

轴测图, axon;

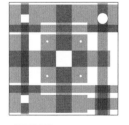

透明性研究, transparency study;

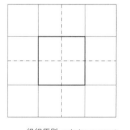

组织原则, ordering concept;

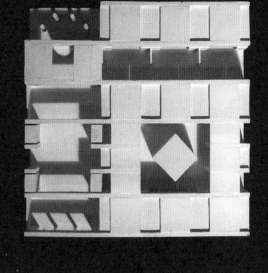

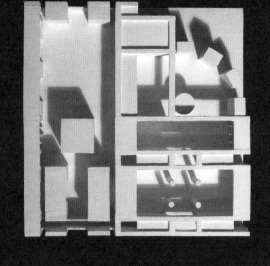

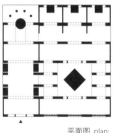

平面图, plan;

立面图, elevation;

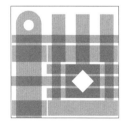

透明性研究, transparency study;

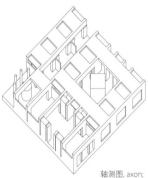

轴测图, axon;

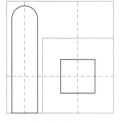

组织原则, ordering concept;

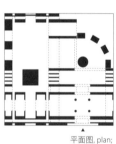

平面图, plan;

立面图, elevation;

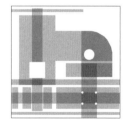

透明性研究, transparency study;

轴测图, axon;

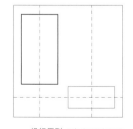

组织原则, ordering concept;

小结 reflection

　　"平行墙"是整套课程中最为综合且富有挑战的一个练习。我们可以将它看作是课程学习的一个"顶点"。在这一设计研究之中，我们试图融汇其他课程练习中所学的知识与方法。该练习讨论的核心主题是"空间限定"，此外还包括诸如流线与路径，密度与虚实关系，水平基准的构成组织，现象的透明性，以及极其重要的设计过程观等问题。对比前面的训练，"平行墙"练习更为复杂，其训练周期亦需要相应延长。在训练中，一个好的设计结果往往需要经历几个发展步骤：首先是最初组织概念的生成，然后利用研究模型来推进平面的深化，通过个人辅导与小组讨论、分析性研究（如空间的透明性图解）以得到最终的设计方案。在完成设计之后，学生需要通过准确且清楚的方式进行方案呈现。描述性的记录包括：表面经白色石膏处理的木质模型，一组包括平、立、剖面与轴测图在内的图纸，以及一套形式图解（如空间透明性研究）。

　　在此，设计过程指的是一种不断试错的工作方法，其中直觉、观察与批判性的分析都十分重要。教学中应该鼓励学生生成一系列可供讨论、比较（甚至

"Parallel Wall" is the most integrated and challenging of the exercises in this program. It is essentially a "capstone" and culminating design study that attempts to bring together all of the lessons encountered in previous exercises. Its central theme is space definition but it also involves important issues such as circulation and path, parti transformation, spatial hierarchy, the role of object elements, density and solid/void relationships, the horizontal datum as a compositional device, phenomenal transparency, and last but not least, the design process. The complexity implied by this exercise requires more time than previous exercises and is best developed through stages that begin with the generation of initial organizational concepts, and are followed by schematically developed plans with study models, individual and group critiques, analytical studies (e.g. spatial transparency diagram) and the final resolution of the design. The last phase requires precision and clarity in presentation. The prescribed graphic documentation consists of a wood model finished in white gesso, a set of drawings including plan, section, elevation, axonometric, and a set of formal diagrams that includes the spatial transparency study.

The design process refers to a working method of trial and error that is informed by intuition, observation and critical analysis. Students are encouraged to generate alternative concepts that can be discussed, compared and

是完全对立）的不同概念，以此开始"自我评价"。设计想法应当及时记录在过程手稿中，这是强调草图重要性的一种方式。学生应当重视徒手草图中比例与尺度的准确性。基于这一考虑，我们在教学中推荐使用无印良品的一款圈装笔记本，其内页纸面印有浅灰色、间隔5毫米的网格点阵。

师生之间的个人辅导可以针对于该学生的设计进行讨论，对学生而言这是一个直接学习的机会。教师有可能直接调整学生的研究模型，比如移动或重新布置其中的部分要素。与此同时，教师可以在具有比例关系的图纸（如平面）上，通过叠加拷贝纸进行草图修改。这不仅是一种有效的教学方式，还可以强调对于设计过程的重视。

练习的起始是一组8片相互平行的墙体，它们等距离排列在一块方形的场地上。如前一练习所讨论的，通过改变墙体之间的间距可以轻易地建立起层级关系。这一操作还有机会创造出密实区域（减少墙体间距离使其紧密排布）与开放区域（增加墙体间距离以扩大空间）。而练习3.4.2"网格构成"中引入的"横纵网格"可以通过以下手段实现：于平行墙上开挖一

contrasted in order to develop a sense of *self-criticism*. Initial ideas are recorded by each student in a journal book that reinforces the importance of sketching and recording ideas visually. An emphasis is placed on developing skill in freehand drawing with correct proportion and scale. The popular Muji@ ring binder notebook with a light 5mm dot screen grid printed on the paper is especially recommended for this purpose.

Individual critiques between instructor and student offer a learning opportunity that is immediate and focused on the student's own work. In this working critique, study models that can be manipulated by the instructor (i.e. elements can be removed or rearranged) are especially effective. At the same time, the use of tracing paper to make revisions by sketching over a scale drawing such as a plan, is not only effective, but serves as a demonstration of design process.

The initial set of eight parallel walls at equal spacing distributed across a square field serves as a point of departure for this exercise. As discovered in the previous exercise, hierarchy can be established easily by repositioning walls at an unequal spacing. This operation has the effect of creating areas of density where the walls are closely spaced, and areas of openness where the walls are separated by some distance. The development of a *cross-grain* grid (as introduced in Ex. 3.4.2 "Grid Composition") is achieved in several ways:

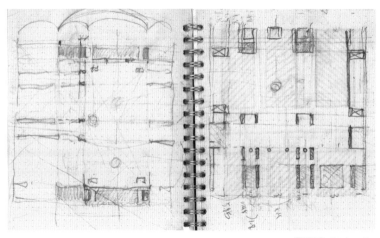

系列相互对位的门洞；引入屋顶板片并使其边缘对齐；引入非结构性墙体；在布置各部件要素时，通过中线对齐或边缘对齐以暗示网格。在此，强调建立网格的目的在于，它可以强化构成组织，在保证整体性的同时允许变形与特例的存在。毫无疑问，网格是一种有效的构成组织工具。在训练中，学生应当充分认识网格的特点、将其发展为一种形式机制，方能创造出丰富且秩序井然的空间构成，否则仍可能制造出一些无趣的设计结果。

练习中有意识地引入了"结构"的概念——承重支撑系统。它同时可作为一种定性的、非强制性的形式机制。基于结构的考虑，我们可以约束一些操作细则，如容许的跨度、洞口上过梁及墙厚。平行墙系统中包括一组厚且相互平行的承重墙体，两种厚度（其设置与跨度无关）的单向屋顶板片。练习中不允许出现悬挑的屋顶板片，墙体可以独立存在（其上不支承屋顶板片）。在更为高阶的平行墙设计训练中，可以引入关于材料与建造方式的讨论。练习中厚墙是主要的结构要素。为避免混淆，我们将4根圆柱视为非结构性部件要素，不鼓励一些企图利用圆柱作为结构支承

creating and aligning openings in the parallel walls; introducing roof planes and aligning their edges; introducing non-structural walls; and positioning object elements to suggest the grid, either through their centerlines or by the object's edges in alignment. The purpose of establishing the grid is that it intensifies the organizational strength of the composition while at the same time, allowing variation and exception to occur. The grid is a compositional tool that can lead to mundane results if not exploited. However, it can also enable the achievement of complexity with order when successfully understood and developed as a formal device.

Structure as a load-bearing support system is introduced in this exercise as a qualitative and non- exacting formal device. Certain rules are given that provide guidelines to span distance, lintel requirements and wall thickness. The system consists of thick, parallel load-bearing walls and one-way spanning flat plate roof elements of two thicknesses (not related to span dimension). No cantilevers are allowed but walls can be freestanding and not support roof planes. More advanced design problems involving the parallel wall system that introduce materiality and construction methods can be proposed. In order not to diffuse the primacy of the thick wall as structure, the round column elements (four) are treated in these exercises as non-structural object elements. Attempts to use the column as an element of struc-

的做法。

在练习3.4.2中首次引入的非结构性墙体（薄的片状要素）在现有的应用中存在多种优势。首先，作为自承重的墙体或分隔，它们可以通过与横纵方向的网格分割线对位，以强化网格的存在。它的另一贡献在于其统一的高度（24mm）与平行墙体上的门洞、部分物件要素的高度吻合，这样就可以创造出强烈的水平基准，暗示剖面上沿垂直方向的划分。这些薄的板片既可以用作独立支承的要素，或是将它们贴附于平行墙墙体的表面，这一做法接近于传统欧洲及西方建筑中所使用的壁板。

类似于结构问题，练习以抽象的方式介绍流线概念——一个运动系统。如练习所要求的，8片平行墙体中有2片位于方形底板的边缘，于其一开设一个门洞作为入口。平行墙上的门洞暗示了通道的存在。这些形式特征可能会激发对于使用、活动与功能计划的直接表达，在这一点上学生可自由发挥。值得注意的是，空间构成不应受到诸如使用、采光、围合等功能性需求的制约。设计中可以假象特定的尺度与使用计划配置，如住宅、图书馆、展厅或是园林集合。抑或是保

tural support should be discouraged.

The non-structural walls (thin planar elements) first introduced in Exercise 3.4.2, have several beneficial applications. As freestanding walls or partitions, they can reinforce a grid by aligning with grid subdivisions in the cross-grain direction. Another important contribution is that their uniform height (24mm) matches the height of openings in the parallel walls and several object elements, thereby creating a strong horizontal datum suggesting a vertical subdivision of the section. The thin planes may be deployed as freestanding elements, or they may be allowed to adhere to the sides of the parallel walls, something akin to wainscoting used in traditional European and western architecture.

Similar to structure, circulation as a movement system is introduced abstractly. Openings in the parallel walls suggest passage and the requirement of an opening in one of the parallel walls positioned on the opposite edges of the square base suggests a point of entry. These formal characteristics may evoke a more literal expression of use, activity and program that the design student is free to associate and work with. While the spatial composition is not influenced by functional requirements such as use, lighting, enclosure, etc., the design may be envisioned for the purpose of scale or configuration as a program for a house, a library, an exhibition gallery, or a garden

持构成本身的抽象性，空间序列是连续的、富有变化与层级关系，保持清晰的形式。

对于设计成果的评价，无需考虑其中与人类活动相关的属性，虽然它可能是学生设计过程的一个起点。评图时应重点审视设计中的形式意图，可针对以下问题讨论：

– 设计是否具有清晰的基本组织结构（构思），它与方形场地的形式属性之间是否存在积极的关系？

– 空间序列是如何配置的？它是否与设计构思存在形式上的关联？其中的各空间形式是否清楚可读且处置到位？对于部件要素的布置，是否在视觉上有助于强化路径与流线模式？

– 设计是否能同时体现加法与减法的空间限定策略？

– 整体上设计是否具有明确的网格结构？其比例是否合宜且发挥组织作用？在配合一些特定的设计特征时，是否容许细微或例外的处理方式？对于网格的进一步划分是否为空间限定与贯通创造出新的机会？

– 设计是否通过使用实体与虚空建立起强烈的图底关系？

complex. Alternatively, the composition can remain abstract, with a spatial sequence that has continuity, variety, hierarchy and formal clarity.

A critique of a design for this exercise need not consider the nature of human activity that may or may not be envisioned by the student. Instead, a review should interrogate the formal intentions of the design. Questions might include:

- Does the basic organizational structure (parti) have clarity and does it relate in a meaningful way to the formal properties of the square field?

- How is the spatial sequence configured? Is it related formally to the parti? Does it involve spatial form that is legible and well articulated? Are the object elements positioned to visually reinforce the path and contribute to the pattern of circulation?

- Does the design incorporate both additive and subtractive strategies of spatial definition?

- Is a grid structure apparent in the scheme? Is it well proportioned and acting as an organizational device? Does it have nuance and exception in accommodating specific features of the design? Do subdivisions of the grid create further opportunities for spatial definition or articulation?

- Is a strong figure ground relationship established through the use of solid and void?

3.5 体验［感知］练习
exercises for **experience [perception]**

3.5.1
剖面空间生成
spatial development in section

任务书 handout

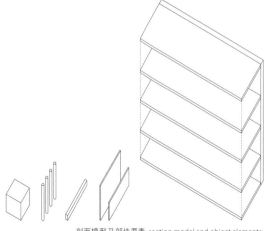

剖面模型及部件要素, section model and object elements;

　　建筑中，剖面是平面的补充。平面与剖面都是通过剖切物件或建筑物获得，其差别在于剖切面的方向不同。为展现三维空间的轮廓，同时表达墙体、楼板、屋顶等建筑要素的厚度，对剖切面的选择应当慎重。通常，可考虑剖切经过建筑中一些重要的开口，如门、窗、天窗等，以便尽可能提供关于这些特别位置的信息。剖切面不应沿与墙面平行的方向剖切墙体，这样会造成一个错觉：剖切到的是一个具有相当深度的实体，而非片状的墙体要素。同样的道理，剖切面不应经过柱子，它会令人误以为剖切到的是一片与剖切面相互垂直的墙体。

　　剖面上的主要建筑要素包括墙体、楼板与屋顶。通过配置这些要素，可以在一个由相互重叠的空间区域构成的紧凑网格构图中，对空间进行限定。这些空间区域可能在大小及重要性上有所差别，它们可用以框定、加强与各物件要素、功能区域之间的对位关系。练习要求根据下述条件进行剖面设计，以创造相互重叠的空间以及空间范围阅读的多样。设计中，各空间应清楚限定，物件要素（立方体、圆柱）可用以强化空间阅读或作为焦点要素。

　　The section in architecture is complementary to the plan. Like the plan, the section is also a cut, made by passing a vertical plane through an object or building. Usually made at a critical location, the section reveals the profile of a space in the third dimension, also showing the thickness of walls, floors and roof. Furthermore, it is preferred that the cutting plane pass through important openings such as doors, windows, skylights, etc. to provide more information about these atypical conditions. The section cut should not pass through a wall plane in the direction parallel to the wall as this will give a false impression of the wall as a deep solid rather than a planar element. For the same reason never cut through a column as it will give the impression that it is a wall plane perpendicular to the cutting plane of the section.

　　The primary elements of a section: walls, floors and roof, are capable of being configured so as to define spaces and spatial fields in a tightly gridded composition of overlapping spatial zones. These spatial zones may vary in size and importance, and may frame or reinforce alignments with objects and functional zones. Within the constraints of the exercise listed below, develop a sectional design with multiple overlapping spaces and spatial field readings. Spaces are to be well-defined and object elements (cube, columns) may be used to strengthen spatial readings or serve as focal elements.

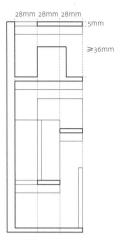

28mm 28mm 28mm
5mm
≥36mm

操作规则, operation limits;

训练要求

　　在一块252mm x 252mm的垂直墙面上，通过组织部件要素来完成一个剖面构成，所使用的墙体与楼板要素如下所述。

　　首先制作垂直墙面，材料为两层厚5mm的瓦楞纸板。这块墙面将成为楼板投形的"底板"。在楼板下表面，可使用截面尺寸为5mm x 10mm的梁对其支撑，以防止出现下垂。楼板共5块：顶层楼板作为屋顶，其上表面需与垂直墙的顶面平齐；底层楼板需紧贴垂直墙体的底面；中间三层楼板的位置可按照设计进行调整，它们需保持与屋顶及地板相平行。设计中，可以将楼板分割成片段，或是挖去一些"洞口"以对位于其他的墙体、部件要素。对于屋顶板，亦可采用类似的挖去操作以将光线引入模型中，如同天窗一般。

　　构成中应巧妙地使用1个立方体与4根圆柱。同时，可考虑使用两种不同高度的分隔墙，需保持其矩形，其上不可开洞口或缺口。在剖面设计中，这些要素可限定空间、创造水平基准线。部件要素如下：

1个立方体	36mm x 36mm x 36mm
4根圆柱	高72mm, 直径5mm

REQUIREMENTS

On a 252mm x 252mm vertical wall plane, configure sectional elements so as to create a sectional composition. Use wall and floor elements as described below.

First create the wall plane (2 layers of 5mm corrugated cardboard) that will act as a "base" for the projecting floor planes. These projecting floors will require beams (5mm x 10mm) attached to the bottom surface to provide support and prevent sagging. There are five floors, a roof plane attached to the upper edge of the wall surface, a ground floor plane attached to the bottom edge of the wall plane, and three floors at a location of your choice between. They must be parallel to the roof and ground floor planes but may be separated into sections and have "holes" cut into them that align with wall elements or objects. The roof plane may also have gaps or holes and these may be thought of as openings for light (sky lights).

A cube and four columns should be strategically employed in the composition. Wall planes of two heights may be used, rectangular with no notches or holes. These will be useful in defining spaces and may also help to create horizontal datums in the section. The elements:

1 cube	36mm x 36mm x 36mm
4 columns	72mm x 5mmø

梁	5mm x 10mm（截面尺寸）	beams	5mm x 10mm (sectional dimensions)
分隔墙	高36mm或72mm，厚2mm	walls	36mm or 72mm in height, 2mm thickness
5层楼板	长度不限，深度需与间距为28mm的垂直参考线对齐（如上图所示）	5 floor planes	length of any dimension, depth must align with the 28mm vertical reference lines (see diagram above)

评价标准

- 满足训练目标与要求
- 整体构成的品质
- 剖面设计中对空间层叠关系的探索
- 空间限定
- 模型制作的准确与手工品质

绘图记录

　　分别选取与垂直墙体平行、垂直的2个剖切面，按照1:1的比例尺，绘制两张剖面图以记录该练习的设计成果。剖切面的选择应当慎重。通常，应剖切经过楼板上的开口以尽可能多地表达信息。同时，绘制两张比例为1:2的分析图解，以阐释设计中空间构成的主要特征，包括主次网格划分与层叠的空间关系。最后将这些徒手绘制的图纸排布在A3白色绘图纸上。

EVALUATION CRITERIA

- Compliance with program objectives and requirements
- Quality of overall composition
- Exploration of overlapping space in a sectional design
- Spatial definition
- Precision and quality of construction

DRAWING DOCUMENTATION

Construct two sections at full scale (1:1). One section should be cut parallel to the back wall plane and the other perpendicular. Choose the best locations for each section. In general, cutting through openings in the floors will reveal the most information. Also draw two diagrams at half scale (1:2) to explain key attributes of the spatial composition of the design. The diagrams should describe major/minor grid subdivisions and overlapping spatial fields. Drawings are freehand pencil on A3 white drawing paper.

SUNG, Chen Ru

LEE, Hiu Yeung Jacky

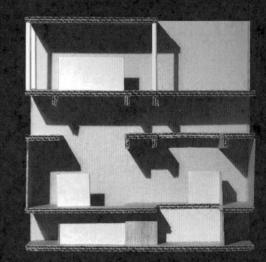

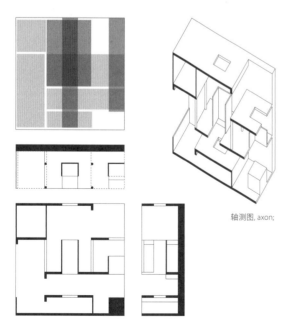

轴测图, axon;

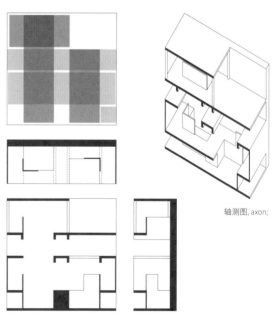

轴测图, axon;

182   剖面及透明性研究, sections and transparency study;

剖面及透明性研究, sections and transparency study;

TING, Wing Lam Phoebe

CHEUNG, Hiu Yan

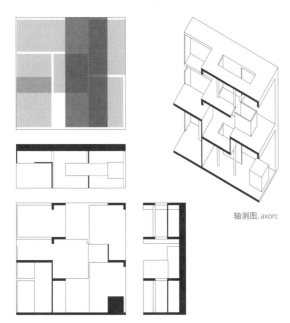

轴测图, axon;

剖面及透明性研究, sections and transparency study;

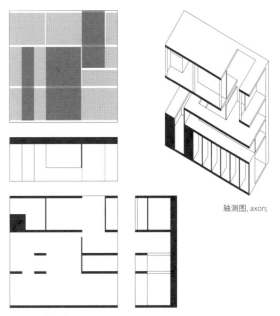

轴测图, axon;

剖面及透明性研究, sections and transparency study;

3.5.2
剖面的氛围品质
ambient qualities of the section

任务书 handout

与平面图一样，剖面图是通过二维抽象的方式来展示三维空间。它在强调空间轮廓的同时，暗示其深度与体积。为了更好地理解设计中创造的空间，我们可以使用纸板（卡纸）来制作比例模型，从中感受空间的体积形式，以及它们之间的关联与分离。然而，为体验模型中的空间，尤其是由光线塑造的氛围品质，我们需要"进入到模型中"。光线可以通过表面反射，或是直接包括空间中的占据物来照亮空间，以此起到模拟光照的效果。

我们通过沉浸的双眼与身体来体验建筑。然而，可视化在模型中获得的体验是极具挑战的。摄影是一种强大而便捷的媒介，它可以将三维的空间与形式直接记录于二维平面上，因而可视为一种有效的学习工具。练习的目的便是要利用照相机来捕捉剖面设计模型中的空间品质，并呈现于一系列的照片中。包括空间层次与深度、咬合的空间体量、形式组织的原则在内的空间特点，光线也应当被记录下来。

训练要求

为拍摄记录设计模型，需搭建一套标准摄影台，

The section drawing, like the plan, reveals three-dimensional space as a two-dimensional abstraction. The profile of space is emphasized while depth and volume is implied. To better understand the spaces of the design, we can build scale models in paper (cardboard) and give a sense of the volumetric form of the spaces as well as their connectivity or separation. However, to experience the space of the model, especially the ambient quality of light, we need to "get inside the model" in order to simulate the effect of lighting that reflects off surfaces or directly envelopes the occupant of the space.

We experience architecture with immersed eyes and body. However, to visualize the experience in the model is challenging. Photography, a powerful and convenient medium for recording three-dimensional spaces and forms directly on a two-dimensional plane, can be a useful tool. The objective of this exercise is to capture the spatial quality of the finished section model through a series of image planes created with a camera. Spatial features such as spatial layering and depth, interlocking spatial volumes, principles of formal organization, and especially lighting, should be documented.

REQUIREMENTS

To create the photographs of the model, use a standard set-up that

185

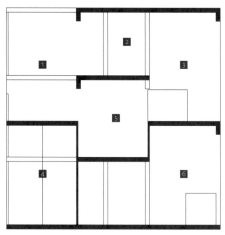

示例, example: TING, Wing Lam Phoebe;

其中包括模型台、射灯、背景布（中性的表面，通常为黑色）以及一部小型照相机。将剖面模型置于模型台上，然后调节射灯的朝向与光线的方向以照亮室内的"房间"。通常，光线应当来自于剖面模型的上方或某一侧边，并使之穿透屋顶板片与楼板上的开口。光与影，明与暗之间的对比，对于获得一个强烈的、三维的表达是至关重要的。

在选取拍摄角度前，你需要充分了解设计的主要特点。练习中，需要拍摄4到6个视角。正面的一点透视是最容易操作的一种视角。拍摄时，应持正照相机，使其正对于模型，以保证模型中水平、垂直布置的要素在照片中处于正确的角度。镜头的高度应当接近于人视点，换句话说，就是空间中的下半部分位置。

拍摄模型的挑战同样来自于对于相机距离的控制：当相机距离过近时，会弱化对于空间深度的表达，模型中靠近相机位置的一些信息可能会被排除于视框范围之外；而过远会将楼板的切面纳入到视框中，消弱于真实空间中的沉浸感。

经过反复的试拍，可从中选取最佳的照片，并使用图像处理软件对照片进行修整，如转化为黑白色彩

includes a model stand, spotlights, background paper (a neutral surface normally black) and a small camera. Place your section model on the model stand and then adjust the orientation and direction of the spotlight to illuminate an interior "room". In general, the light should come from above or either side of the section model, and penetrate through the subtractions in the roof and floor planes. The contrast between light and shadow, bright and dark is essential to achieve a strong and three-dimensional expression.

Familiarise yourself with the design features before selecting the viewpoints. In this exercise create from four to six views. The frontal one-point perspective is the most manageable view. You should hold the camera perpendicular to the model to ensure the horizontally and vertically placed elements are at right angles in the photo. The lens should be placed at a similar level as the eye, in the other words, in the lower part of a space.

The challenge of photographing this model also comes from the control of camera distance. Too close and the presentation of spatial depth will be weakened and certain information in the foreground will be eliminated. Too far away and the picture area will include the sectional edges of the floors, foiling the sense of being immersed in an actual space.

After several trials select the best photos and retouch them with Adobe Photoshop, i.e., convert into a black/white mode, adjust brightness and

模式、调整明暗对比等。照片的比例可以根据排版需要进行调整，最后将黑白照片打印出来，将其裁剪后根据排版黏在厚的A3白纸上。

评价标准

– 视角的选择与空间概念的表达
– 光线环境，调整以实现空间的最佳表现
– 对于透视法则的理解
– 照片的品质，如照片的分辨率与清晰度，黑白对比，构图等

darkness contrast, etc. The proportions of each photo may be adjusted to fit the A3 panel layout (white thick paper), on which the black and white photocopies are mounted.

EVALUATION CRITERIA

-   Selection of views to communicate spatial concepts
-   Lighting conditions: adjusted to provide the best depiction of space
-   Acknowledgement of the laws of perspective
-   Quality of image, e.g. image resolution and clarity, black/white contrast, composition, etc.

小结 reflection

课程的最后一组练习，突出强调使用者的空间体验。这里，我们开始更加直接地讨论人体与空间尺度的关系，以及欣赏包括光影效果在内的空间氛围品质问题。

到目前为止，课程的大部分练习都可以认为是平面主导，其中对于空间的塑造与配置主要是通过平面的拉升来实现。根据墙体与门洞的高度，可以初略估计其人体尺度。事实上，空间的准确尺寸对于其配置及空间概念的影响甚微（各空间的比例是一个与之无关的问题）。前面各构成练习相对重视比例的序列关系、空间或要素之间的关系（如对称、层级性、轴对称性等）以及各要素之于场地的关系。设计要求在"计划"所设定的形式限制（如协调特定的物件要素）与场地的形式关系（无方向或周围环境关系的抽象方形底板）之间寻求平衡。从外形上，我们可以从形式的角度来欣赏这样的构成，并倾向于使用轴测的图示方式进行方案呈现。

考虑到人体尺度的问题，真实的占据空间会带来一系列影响使用者感知的新问题。首先是空间尺度或尺寸与人体的关系。一个空间或者是一个要素（如门

The final series of exercises in this program emphasize the *spatial experience* of the user. Here we begin to introduce more directly the scale of a space in relation to a person, as well as an appreciation for the ambient quality of space, such as the effect of lighting.

Up to this point, most of the exercises may be described as *plan active*; spaces are formed and configured primarily as extrusions of the plan. Human scale can be surmised based on wall heights and openings but the exact dimensions of spaces have little impact on the configuration and spatial concept (the *proportions* of the spaces are an entirely different matter). Composition until now is a consequence of proportions, relationships between spaces or elements (symmetry, hierarchy, axiality, etc.), and relationships of elements to the geometry of the field. The design is a balance between formal constraints of the "program" (such as the accommodation of specific object elements) and a formal relationship to the "site" (the abstraction of a square base with no directionality or context). Such a composition can be appreciated externally as a form, with the preferred view an axonometric.

With human scale considered, the reality of occupying space introduces new issues impacting the perception of the user. First is the scale or dimension of the space in relation to a person. The appropriate size of a space or

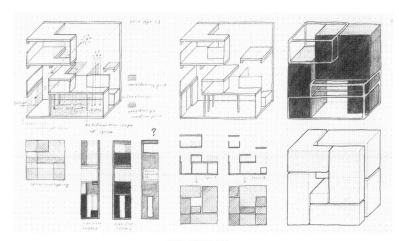

设计过程图解分析, sketch record and diagrams (WU, Yu Shang Sunny);

洞或窗户）的尺寸通常取决于具体的使用功能。为满足通行的需要，门洞必须有足够的高度，但又无需过高。同样重要的另一问题是"视点"，即空间中人眼的位置。在透视图中，人眼高度决定了图纸建构时视平线的位置。这就变成了另一种基准：视野之中的物件将处于视平线以上或以下。同时，透视图有利于我们从使用者的视角来理解室内空间。

接下来应该考虑的是"文脉"——设计所处场地及其周围环境的特点。自然的作用是场地环境文脉中的一部分，它们会对设计产生影响。在这个练习中，学生应当考虑自然光的影响。自然光通过包括屋顶在内的外部包裹进入到空间之中。为进一步讨论空间限定，我们在训练中将平行墙模型旋转至垂直方向，这样墙体便变成了水平的楼板，而最上方的板片便是屋顶。平行墙练习要求在一尽端的平行墙体上开挖门洞作为场地的入口；而在此练习中，屋顶板片上的洞口是为了引入自然光线。

练习3.5.1的重点在于：通过对水平板片进行减法操作（移除部分楼板以创造楼层之间的开口）以及置入物件要素与非结构性墙体板片的加法操作，来创造

an element (such as a doorway or window) is partially determined by function. The door opening is high enough to afford passage without being too tall. Equally important is the *viewpoint*, or position of the eye in space. In a perspective drawing the height of the eye determines the position of the horizon in the construction. This becomes a type of datum in which everything in the view is either above or below the horizon line. The perspective helps us to understand the interior space from the point of view of the user.

Next one may consider the *context*, the characteristics of the surrounding area and the place where the design is located. The forces of nature are part of the environmental context of the site and they have an impact on the design. In this exercise natural light is considered. Natural light enters the space through the exterior envelope that includes the roof. To explore space definition in this exercise we imagine the parallel wall exercise rotated into a vertical position. The walls are now seen as floor planes with the uppermost plane as the roof. As in the parallel wall exercise, the end plane must have an opening; here, instead of an entrance, it is an opening in the roof for light to enter.

Exercise 3.5.1 is focused on the spatial conditions that may be created by using the *subtractive* process on the floor planes (removing portions of the floor to create openings between floors) and the *additive* process of intro-

多种多样的空间状态。减法的操作方式可以解决单一楼层高度的限制，并创造穿越多个楼层的空间。通过改变楼层高度，同样可以加强空间尺寸上的对比（如剖面上的大、小开间）。而物件要素（1个立方体、4根圆柱与薄墙板片）的布置将强化空间限定，同时在剖面上暗示深度的不同层次（模型沿剖面的深度方向被划分为3个厚度为28mm的区域）。

为加强水平楼板的刚度，练习引入了截面尺寸为5mm x 10mm的梁要素，它可以布置在楼板的下方，与背面墙体保持相互垂直。除了为悬挑的楼板提供有效的结构支撑外，这些梁还是一种重要的片状要素，它可以为设计中一些空间体量的边角提供额外的限定。或是强化了剖面网格中那些打破楼板连续性的垂直空间带。这些高度为10mm的梁同样可以定义次一级的水平基准，并通过物件要素或片墙的布置予以强化。

教学中，应指导学生通过运用这些操作方式来创造一个具有丰富空间的剖面；在保持整体构成策略控制的前提下，使用层级关系、空间区域与空间的层叠来获得变化。

在练习3.5.2中，学生需要使用一种新的工具——

ducing object elements and non-structural wall planes. The first operation, subtraction, overcomes the limitation of the single floor height and creates spaces that extend through several floors. Floor heights may also be varied to increase the contrast in size (big bay, small bay in section). Placement of the object elements (a cube, 4 columns and thin wall planes) helps to increase the space definition as well as suggesting layers of depth in the section (the subdivision of sectional depth into three implied zones each 28mm in width).

To provide stiffness for the projecting floor planes, the exercise introduces beam elements (5mm x 10mm) that may be placed underneath the floor planes and perpendicular to the back wall. Aside from the useful structural support they give to the cantilevering floor planes, the beams are important planar elements that give additional definition to the corners of spatial volumes in the design. They reinforce vertical zones in the grid of the section interrupting the continuity of the floor planes. The 10mm depth of the beams can also define secondary horizontal datums that may be reinforced by the height of objects or thin wall planes.

Students are directed to use these operations to create a section that has spatial complexity, using hierarchy, spatial zoning and overlapping spaces to achieve variation within the overall compositional strategy.

评图准备, review pin-up;

相机来对练习3.5.1的设计成果进行记录。剖面模型的设计为此创造了条件：允许相机足够靠近模型，并浸入式地捕捉室内空间。拍摄时，取景框与剖面的深度方向相互垂直（也就是与背面的墙体相平行），以此可获得空间的一点透视。在拍照记录的过程中，学生需要探索自然光的影响，使其从模型的两侧边及屋顶上的开口进入到剖面空间中。由此可以观察直射光与非直射光的不同作用效果。在拍摄时，应注意取景框中的视平线与比例模型空间中人眼位置间的契合。假设剖面模型的比例尺为1:40，那么楼板之间的平均净高约为2.5m，通常来说人眼高度离地约1.6m，换算后相机镜头需距离模型中各楼面约40mm。按照这一方式拍摄，便开始了真实空间氛围的复制。

In Exercise 3.5.2 students document the completed design of 3.5.1 using a new tool: the camera. The model is designed in such a way that it is possible to obtain good close-up photographs of the interior spaces. These photos are made perpendicular to the depth of the section (that is, parallel to the back wall) and thus resemble a single point perspective of the space. In conjunction with the documentation, students explore the effect of natural light entering the section through the two sides of the model as well as from the opening in the roof plane. The effect of direct and indirect lighting may be observed. In conjunction with the implied horizon line coinciding with the eye of the inhabitant (assuming a scale of 1:40 translates to an average floor to floor height of about 2.5m, and the height of the eye of the observer is about 1.60m or 40mm above the floor) the photographed space begins to replicate the ambience of an actual space.

date		studio activity	assignments due	reading
T	1/8	Introduction		PROCESS
Th	1/10	EX. 1 FREEHAND LINE		TECHNIQUE
T	1/15	EX. 2 DRAWING STUDY		PRINCIPLE
Th	1/17	EX. 3 SINGLE VOLUME		Ching:Arc pp.331-38
T	1/22	REVIEW EX. 3	HW 1 Due	COMPOSITI
Th	1/24	EX. 4 SQUARE COMPOSITION		Arnheim:A pp.10-41,
T	1/29	EX. 5 SPACE DEFINITION	HW 2 Due	ANALYSIS
Th	1/31	REVIEW EX.4 & 5		ANALYSIS
T	2/5	EX. 6 PARALLEL WALL I Introduce PROJECT	HW 3 Due	GRID
Th	2/7	EX. 7 PARALLEL WALL II		GRID
T	2/12	REVIEW EX.6 & 7,& PROJECT	HW 4 Due	SPACE
Th	2/14	EX. 8 SPATIAL TRANSPARENCY		TRANSPARE
T	2/19	REVIEW PROJECT	Model & Plan	
Th	2/21	EX. 9 OBJECT WALL		
T	2/26	EX.10 DATUM WALL		
Th	2/28	REVIEW EX.10	HW 5 Due	
T	3/5	REVIEW PROJECT	Final Study Model & Plan	
Th	3/7	Model Construction	Final Drawings Due	

4 回探

backdrop

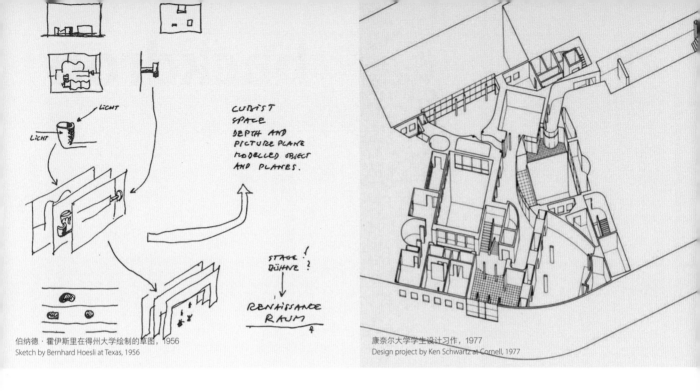

伯纳德·霍伊斯里在得州大学绘制的草图，1956
Sketch by Bernhard Hoesli at Texas, 1956

康奈尔大学学生设计习作，1977
Design project by Ken Schwartz at Cornell, 1977

布鲁斯·朗曼 Bruce Lonnman

4.1 溯源 path

形式主义教学思想的发展 origins of a formalist pedagogy

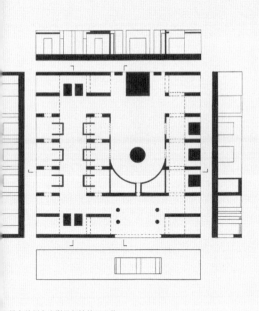

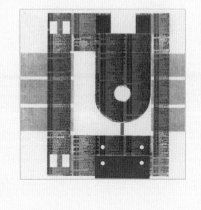

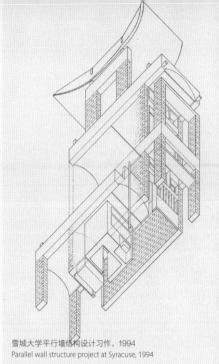

俄亥俄州立大学平行墙练习习作
Parallel wall project by Deb Shumaker at Ohio

雪城大学平行墙结构设计习作，1994
Parallel wall structure project at Syracuse, 1994

可以说，建筑设计入门的教学方法是一个相对新的事物。在包豪斯之前，对于建筑学基础的学习主要是通过在职业建筑师的事务所中担任实习生来实现。其后巴黎美院建立了一个相对结构化的培养方式，它同时结合了图房训练与相关的竞赛机制。在20世纪中叶，有关设计的专业教育开始以一种更为结构化与系统化的方式来讨论建筑学的教学问题。其中的一种方式主要来自于1954至1959年的得州大学奥斯汀分校，参与其中的教育者们被后人称作"得州骑警"。他们制定的教学大纲通过识别现代建筑的风格特征，揭示其潜在的、无关建筑所处历史时期的形式品质来引入一种对现代建筑更为严格及批判的视角。这一方式的重要特点在于对空间设计的强调及其发展出的一系列抽象空间设计练习。

本文追溯了作者对于这一教学方法的最初认识，以及其后以建筑学中空间设计为主体生发出的一套基础设计教案。

The pedagogy of beginning architectural design teaching is a relatively recent development. Prior to the Bauhaus, learning the fundamentals of architecture was typically an internship in the office of a practicing architect. Then the Beaux-Arts established a more structured program with both *atelier* training and related *concours*. By mid-twentieth century, academic programs in design began to address the teaching of architecture in a more structured and systematic way. One such approach originated at the University of Texas in Austin from 1954-59. The educators involved were later referred to as the "Texas Rangers". The curriculum they shaped introduced a more rigorous and critical view of modern architecture, identifying its stylistic attributes and revealing its underlying formal qualities inherent in all periods of architecture. An important characteristic of the approach was its emphasis on spatial design and the introduction of abstract and formal design exercises.

The essay *Path* traces the author's initial exposure to this pedagogy and the subsequent development of a program for teaching basic design based on the primacy of spatial design in architecture.

第一节：得州骑警｜得州大学 1953-1958

　　1971年，康奈尔大学建筑艺术与规划学院的一名年轻助理教授罗杰·舍伍德开设一门面向建筑系新生的历史/理论入门课程。其重点在于对过往建筑与城市案例的形式分析。从古埃及遗迹一直到20世纪中期的当代建筑，该课程涉及的案例广泛。但是，这并不是一门有关历史调研的课程，选取的案例旨在图示关于秩序的形式概念：轴线、对称、层级、基准与重复。配合课堂讲座的课程讲义是一系列图版［图4.1-1］，主要以平面、剖面、立面及轴测图为主（有少量的透视图），并在边栏中对于展示案例与五原则之间的关系进行简要注解。其后，这本名为《建筑的原则与要素》的讲义由南加利福尼亚大学出版。它对于《建筑：形式、空间和秩序》[1]（设计入门相关的重要著作之一）的作者程大锦产生了很大影响。

　　在康奈尔大学，有一群被称作是"得州骑警"[2]的教师，而罗杰·舍伍德正是其中的一员。得州骑警来自德州一所特别突出的大学——得州大学奥斯汀分校（后文简称得大），它所在城市与康奈尔相距甚远。当时是20世纪50年代中期，这个学校里有一帮年轻老师，他们虽没有很多教学经验，但怀揣着一个很大的设想：创造一种全新的建筑教学方法，它既能识别、吸收过往的建筑传统，也能对于当下的建筑趋势（尤其是现代建筑）进行批判性的分析与应用。

　　起初，令人意外的是他们采用了一种相当直接的方式：分析现代建筑大师的重要作品，以发现其中的

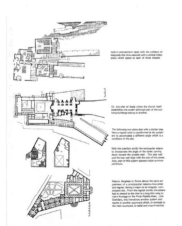

图 4.1-1

《建筑的原则与要素》，内页图版；

Principles and Elements of Architecture, plates;

Part 1: The Texas Rangers｜University of Texas 1953 -1958

In 1971, Roger Sherwood, a young Assistant Professor in the College of Architecture Art and Planning at Cornell University taught an introductory history/theory course to freshmen in architecture. His emphasis was on the formal analysis of works of architecture and urban design of the past. The survey was broadly inclusive with examples from ancient Egypt up through the contemporary architecture of the mid-twentieth century. But this was not a historical survey course and the precedents were selected to illustrate formal concepts of order: axis, symmetry, hierarchy, datum and repetition. The class notes that accompanied his slide lectures were a series of plates [fig. 4.1-1] containing images; mostly plans, sections, elevations and axonometrics (a few perspective images were included) with brief sidebar notes to succinctly explain the formal attributes of the illustrated work in relation to the five principles. These notes were later published as *Principles and Elements of Architecture* (University of Southern California) and influenced Francis D. K. Ching, author of *Architecture: Form Space and Order*, one of the leading texts on Introductory Design[1].

Roger Sherwood was a member of a group of architecture faculty at Cornell University who came to be known as the "Texas Rangers"[2]. The genesis of the Texas Rangers was far distant from Cornell at a prominent university in Texas (the University of Texas in Austin, henceforth UT). There in the mid-fifties a few young and relatively inexperienced teachers at the school shared a bold vision, to invent a new method of teaching architecture that would recognize and incorporate the architectural traditions of the past, while at the same time, be critical in the analysis and application of current tendencies (especially that of Modern architecture).

At the start, they adopted a surprisingly direct approach; analyze the important works of the leading architects of Modern Architecture to discover their guiding principles and formal characteristics. Later, both contemporary and historical architecture were subjected to critical study in order to discern commonalities, that is, formal characteristics that are enduring and evident in all significant works of architecture. It was assumed that modern architecture was part of a tradition while, at the same time, possessed specific formal elements that identified it as a style.

The initial research of Bernard Hoesli and Colin Rowe, the two main protagonists of the Texas Rangers, focused on works of the leading masters of the era, namely Frank Lloyd Wright, Mies Van der Rohe, and especially, Le Corbusier. Charles Jeanneret, aka Le Corbusier, was acclaimed as a key originator of Modern Architecture but was not so well known in the USA. His work was heavily European and nuanced by political and cultural exigencies that

关键原则与形式特征。之后，他们对当代建筑与历史建筑进行批判性的研究，认识他们之间的共性，即所有重要建筑作品中经久不衰的、显著的形式特征。他们假设现代建筑设计是传统中的一部分，与此同时，识别其中一些特定的形式要素为一种风格。

在得州骑警里，伯纳德·霍伊斯里与科林·罗是关键的两名主角。他们最初的研究关注于当时一些重要的建筑师，如弗兰克·赖特、密斯·凡·德罗以及勒·柯布西耶。柯布西耶的原名是查理斯·乔纳雷，他被认为是现代建筑的关键源起，但是在美国却不为人熟知。其作品带有强烈的欧洲风格，并受当时的政治文化氛围影响，这一点也是北美观众所不熟悉的。然而，柯布作品中最关键的是其形式严格，这亦肯定了建筑构成的基本原理。之后出版的《柯布西耶全集》一共7卷，按照时间顺序详细记录了各方案的设计过程及其理论思想。以此，他的作品逐渐广为人知。

霍伊斯里与科林·罗的这些研究也使得他们思考一种原创性的建筑教育方式：重视对于建筑历史的研究，将从研究中获得的启示转译到课程之中，为设计教学提供一种有序且目标明确的方法。写于1954年的一份"宣言"表述了他们所确信的观点："现代建筑设计是可教的；在所谓的现代主义运动中，存在大量重要的建筑项目；现在完全有可能来分析理解这些案例，并从中抽取出一套行之有效的建筑理论。"3

这篇文章继而甄别了4个重要的课程目标："（1）本质上，设计过程是对一个给定情况的批判；（2）应当唤醒（学生内心）概括与抽象的力量；（3）筛选这一举动假设了对特定原则的遵从；（4）一所教育机构应当提供基本的知识与态度。"4

最终，因内部分歧而出现的一系列事件导致了这批激进的年轻教员的辞职、离开。霍伊斯里返回了位于苏黎世的瑞士联邦理工学院。科林·罗在剑桥大学停留了些许年后，开始了在康奈尔的长期教职生涯，约翰·肖、李·祝辰以及沃纳·塞利格曼也随后加入。约翰·海杜克与罗伯特·斯拉茨基则接受了在纽约库伯联盟的教职。这样，发迹于得州的这一激进建筑教育改革现已经转移到了美国的东海岸，同时也跨海去到瑞士。

在得州的时间虽短，科林·罗、霍伊斯里等领导的教学改革还是为空间设计引入了一些重要概念，并

were unfamiliar to a North American audience. Yet underlying all of Le Corbusier's work is a formal rigor that conforms to the basic fundamentals of architecture composition. Furthermore, the work of Le Corbusier became widely disseminated through the influential publication of the Ouevre Complete, a seven volume serial edition that detailed his design process and theoretical ideas within a chronological presentation of his work.

The research of Hoesli and Rowe led them to propose an original approach to architectural education that embraced the study of architectural history and translated the insights gained from the research into a curriculum that offered a sequential and objective approach to design education. A 'manifesto' written in 1954 expressed the conviction that "modern architectural design, could be taught; that there existed a large number of significant buildings and projects within the so-called modern movement; and that it was now possible to analyze these examples, and to understand and extract from them a workable, useful body of architectural theory." 3

The paper went on to identify four important objectives of the curriculum. These were: "(1) that the process of design was essentially the criticism of a given situation, (2) that the power of generalization and abstraction (in the student) must be aroused, (3) that the act of selection assumes a commitment to certain principles, and (4) that an institution should offer an essential knowledge and an essential attitude." 4

Eventually a series of events attributed to departmental politics led to the termination and then departure of the younger and progressive faculty members. Bernard Hoesli had returned to the Swiss Federal Institute of Technology (the ETH) in Switzerland. Colin Rowe, after a few years at Cambridge University, began a long teaching career at Cornell University and was joined by John Shaw, Lee Hodgden, and Werner Seligmann. John Hejduk and Robert Slutzky accepted positions at the Cooper Union School of Architecture in NYC. Now the radical reforms in architectural education initiated in Texas shifted to the east coast of the US and overseas, to the ETH.

During those brief few years in Texas, the pedagogical reforms created by Hoesli, Rowe, et al, introduced several important concepts pertaining to spatial design and were tested in design exercises created for the teaching program. Chief among these was a conception of space referred to as phenomenal transparency.4 Mainly associated with the work of Le Corbusier, phenomenal transparency seemed to distinguish his work from the other moderns. [fig. 4.1-2] While the *free plan* suggested a conception of space as a continuum, phenomenal transparency implied a spatial sensibility that comprehended space as form. Terms like spatial layering, deep and shallow space, spatial fields and zones described a perception of space that has

在课程的设计练习中进行试验。其中最主要的一个空间概念是"现象的透明"[4]。这一概念通常与柯布西耶的作品联系在一起，并以此与其他现代建筑大师的作品相区分。[图4.1-2] 当"自由平面"视空间为一个连续体，"现象的透明"则暗示了一种将空间理解为形式的空间敏感性。诸如空间层次、深空间与浅空间、空间场域与区域在内的术语描述了一种对具有尺寸、边界，甚至形状的空间的感知方式。这种视空间形式为一种建筑设计关键要素的观念与当代作家、评论家布鲁诺·赛维的思想不谋而合。赛维在其著作《建筑作为空间》[6]中曾指出，空间塑造是建筑设计的首要目标。通过分析米开朗基罗所设计的罗马圣彼得教堂以及其他案例，赛维将注意力转移到对于空间图形的描绘上。

得大的新课程计划带出了一种新式的建筑教学方法，其中包括一些限制明确的抽象形式设计练习，以此强化讲座中所讨论的空间的层次与接合。这些说教性的"练习曲"（如"入门练习4"——在方形网格上使用立柱、隔墙与水平板来创造一个包括4个空间的序列 [图4.1-3]）挑战了学生对于空间设计，而非占据空间或塑造空间的实体要素的关注。大部分练习是平面性的，针对二维的构成问题。

这当中有一项非常重要的教学法成果——对于网格化空间区域的探索，这导致了经典练习"九宫格问题"的出现。其后，约翰·海杜克在练习的基础上构

dimension, boundary and even shape. This emphasis on spatial form as a key element of architectural design had affinity with the thought of contemporary writer and critic, Bruno Zevi, who wrote the widely read book, *Architecture as Space*[6] in which he stated that the shaping of space was the primary objective of architecture. Zevi brought attention to the depiction of space as *figure* in his analysis of Michelangelo's design for Cathedral St. Peter's in Rome, as well as other examples.

The new curriculum at UT introduced a method that was relatively new in architectural education involving limited and abstract formal design exercises intended to reinforce lessons about spatial layering and the articulation of space. Many of these didactic *études* such as "Introductory Exercise 4" (create a sequence of four spaces in a square grid with columns, partitions and horizontal planes) [fig. 4.1-3], challenged students to focus on the design of space rather than the solids that occupy space or that form the space. Most of these exercises were plan oriented and dealt with two dimension compositional issues.

One of the more significant outcomes pedagogically was the exploration of gridded spatial fields that led to the iconic nine square problem. First imagined by John Hejduk and related to his personal design studies of hypothetical nine-square grid houses[7] [fig. 4.1-4], the nine-square problem later became a rite of passage in the first year syllabus of the Cooper Union. In short, the nine-square was a grid of 16 equally spaced columns creating a uniform and neutral field within which the design, incorporating solid elements of specified dimensions and shape, would be formulated. [fig. 4.1-5] Abstract and precise, the nine-square problem encouraged a non-classical

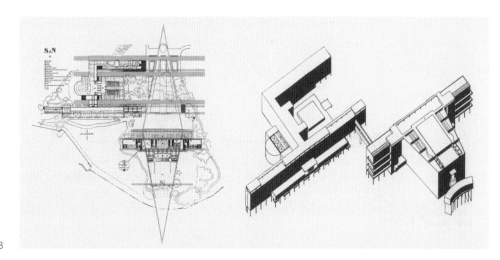

图 4.1-2

现象的透明示意：国际联盟总部，通过建筑体量对位暗示平面上相互平行的空间带；

Example of phenomenal transparency: League of Nations, plan with parallelel spatial zones implied by the alignment of the edges of the building mass;

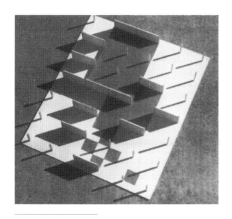

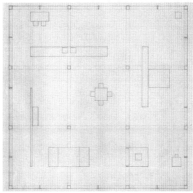

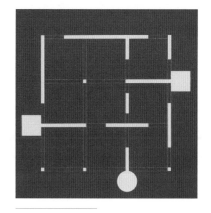

图 4.1-3
得州大学"入门练习4"学生习作模型；
Study model of "Introductory Exercise 4" at UT;

图 4.1-4
住宅五平面，约翰·海杜克，1960-62；
Plan of House 5 by John Hejduk, 1960-62;

图 4.1-5
九宫格练习学生习作，库伯联盟；
Student work of nine-square exercise at Cooper Union;

想了一系列九宫格住宅的设计研究[7] [图4.1-4]，并将该练习作为库伯联盟一年级教学大纲中的入门训练。简单来说，九宫格是16根等距均匀布置的网格，它创造了一个统一、匀质的场地，通过置入特定尺寸与形状的实体要素来生成设计方案。[图4.1-5]"九宫格问题"是一个既抽象又精准的练习，它期望塑造关于构图的一种非经典的感觉：接受非对称、暗示性的向心运动，不完整的格式塔图形以及一个无边界的空间连续体。这一练习之后不断成熟，学生需要根据要求自行制作各要素部件。这一组木质要素便是设计的工具箱，又称装配部件。工具箱里的这些几何实体要素会在复杂程度递增的多个形式设计练习中反复使用，并以此讨论更为广泛的构成与空间问题。

第二节：白色建筑 | 康奈尔大学 1963-1975

科林·罗于1963年加入康奈尔大学，不久后便建立了一个全新的研究生课程——城市设计。加上他之前在得州大学的同事（包括约翰·肖、李·祝辰、沃纳·塞利格曼），这群"得州骑警"对康奈尔的设计教学产生了直接的影响。科林·罗将他大部分的时间用在了这个城市设计课程上，并特别发展了针对城市设计的形式主义方法。而肖、祝辰、塞利格曼则主要教授五年制的本科课程。在塞利格曼的带领下，这一课程逐步成为北美最严谨、正统的建筑学教学大纲之

sensibility towards composition that embraced asymmetry, implied centripetal movement, incomplete gestalt figures and an unbounded spatial continuum. As the design exercise matured in time, students were required to manufacture their own elements according to specifications. The set of these wood components was a design toolbox referred to as the "kit of parts". This toolbox of solid geometrical components served a number of formal design exercises of increasing complexity that addressed a wide range of compositional and spatial issues.

Part 2: White Architecture | Cornell University 1963-1975

Soon after Colin Rowe was appointed to the faculty of Cornell University in 1963, he established a new postgraduate program in Urban Design (UD). Reunited with former University of Texas colleagues John Shaw, Lee Hodgden and Werner Seligmann, the so-called "Texas Rangers" had an immediate impact on design teaching at Cornell. While Rowe devoted most of his time to the new UD program developing a unique formalist approach to the design of cities, Shaw, Hodgden and Seligmann taught primarily in the undergraduate 5-year professional degree program. Under the leadership of Werner Seligmann this program quickly established itself as one of the most rigorous and formal architecture curriculums in North America. The influence of Rowe's unorthodox but inspired approach to history permeated the culture of Cornell and the unpublished (at the time) essays[8] together with his brilliant lectures on Mannerism and Modern Architecture established the theoretical underpinning to the pedagogy of the program.

一。科林·罗对于历史所抱持的非常规但极具启发性的方式影响并渗透了康奈尔的文化。他针对手法主义及现代建筑的一些写作（当时尚未出版）[8]与演讲为这套课程建立了理论基础。

塞利格曼教授的一年级设计课程中引入了诸多当年他们在得大发展出的形式设计练习。配合二维与三维模型的抽象空间限定练习的是对于校内建筑室内房间的徒手绘图分析研究。[图4.1-6]这些建筑是典型的传统风格（都铎、新古典主义、罗马式等），在练习中它们犹如一个个足尺模型，学生可以以此学习基本的平面、立面、轴测图的建筑制图技巧。与此同时，这些房间有着较好的空间处理，可用于深入的形式分析。最后一套完整的成果图纸包括平面、剖面、内立面、剖切轴测图以及表现空间分割、水平基准、房间内表面构成的形式分析图解。与此同时，图纸上还会记录一些功能方面的信息，包括流线模式、家具布置、灯光等。这些图需绘制在8½英尺 x 11英尺（21.59cm x 27.94cm）的纸上，然后依次首尾相连地黏在一起并折叠成一本本如手风琴的小册子，又称莱波雷洛装帧法[9]。

平行于设计课的是舍伍德开设的"建筑的原则与要素"，课程旨在介绍建筑的基本形式概念。舍伍德选择了一些代表性的案例，并通过图解分析来展示包括轴线、对称、层级等在内的形式组织原则。

舍伍德于1973年离开康奈尔，约翰·肖所带领的建筑学基础课程逐步向"设计"靠近。每一堂课都有

In the first year design studio led by Seligmann, students were introduced to many of the same formal exercises that were developed a decade earlier in Texas. The abstract space definition exercises in 2D and 3D models were complemented by analytical freehand drawing studies of existing rooms in buildings on the campus. [fig. 4.1-6] These were typically of traditional style (Tudor, Neoclassical, Romanesque, etc.) and served as full-scale models for learning the basic architectural drawing techniques of plan, section and axonometric. At the same time, the rooms were spatially well articulated and capable of sustaining an in-depth formal analysis. A complete set of drawings would include plans, sections, interior elevations, cutaway axonometrics, as well as formal diagrams expressing spatial subdivisions, horizontal datums, and compositional analyses of interior surfaces. Certain functional considerations such as circulation patterns, furniture arrangements, and lighting were also recorded. The drawings made on individual sheets of paper (8 ½ x 11 inches) were then taped together end-to-end forming a folded booklet like an accordion and referred to as a *Leporello*[9].

Parallel to the studio, Sherwood's complimentary course in *Principles and Elements of Architecture* offered a primer on basic formal concepts in architecture illustrated with diagrammatic analyses of historical precedents chosen to illustrate the particular concept (e.g. axis, symmetry, hierarchy, etc.).

After the departure of Sherwood in 1973, the course in fundamentals evolved under John Shaw as a more design oriented approach. Each class period introduced a brief (an instruction with objectives) that described a type of space study exercise explored in simple cardboard models. A typical brief: "With 25 – 1 inch high vertical planes of any length define four spaces,

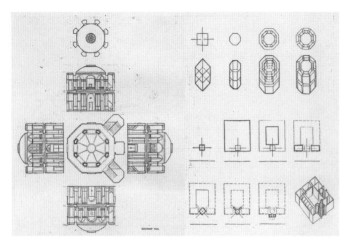

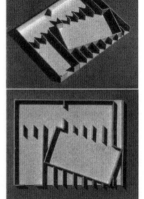

图 4.1-6
房间图解分析示例: 沙利文特堂;
Example of a room drawing analytique: Sullivant Hall;

图 4.1-7
空间限定练习（8 ½英尺 x 11英尺底板）示例;
Example of a space definition exercise (8 ½ x 11" base);

一份任务书（列明教学目标的指导书），其中布置了需使用简单卡纸模型来进行空间研究的设计练习。典型的任务书是这样描述的："使用25片高1英尺（2.54cm）、长度不限的垂直纸板来限定四个均为矩形的空间，且其中一个处于层级关系中的主导。承放这些墙体的矩形卡纸底板大小为8½英尺 x 11英尺（21.59cm x 27.94cm）。"[图4.1-7] 这些抽象练习预期在短时间内完成（如两小时），并聚焦于某一特定的问题，即"每日一题"。

可能一年级的抽象设计练习中最具挑战的是"立方体问题"。任务书的要求很直接：使用卡纸（厚2毫米的斯特拉斯莫尔白色展览用纸板）来建构一个边长为5英尺（12.7cm）的立方体。[图4.1-8] 在这个立方体中置入三片沿立方体正交方向布置的板片，以此限定xy，yz与zx平面。立方体外表面的板片应当是矩形，且不完全覆盖该立方体的各表面。这个练习对于"准确性"有着较高要求，模型制作时应注意板片的切割与黏接。虽然练习强调培养学生的模型制作能力，但这并不是该练习的主要目标。练习的根本目的在于强化空间感知。立方体相平行的两表面上的板片边缘与内部的三块矩形板片边缘相互对齐，可使得一系列清楚限定的长方体体量相互接连。这里面的另一要点是建立空间的层级性。这一想法是为了避免出现"棋盘格式"的毫无层级差异的空间限定。这也自然导向了非对称的板片以及其相应空间体量的配置。

"立方体问题"逐渐成为康奈尔一年级设计工作室的通行证。塞利格曼是这个练习的拥护者，也有可能他就是这个练习的设计者。他相信这个练习可以强化对于空间的敏感性，让学生在一个真实的设计语境中提升创造与操作空间的能力。"立方体问题"挑战了设计者立方体块分割，在立方体边界范围内从剖面的角度来组织一系列的空间体量。另外，表面板片对于立方体的立面构成亦起着重要作用。它可以通过限定内部空间体量，阅读建筑的内部空间来强化立面的现代主义概念。

在学生所经历的不同设计课程训练中，先例的重要意义得到了反复强调，并在不同层次上发挥作用。先例的一种使用方式是介绍"建筑类型"概念，它与功能计划或使用有关。先例也可以用来讨论各种形式类型，比如说包含庭院的居住建筑。值得注意的是，

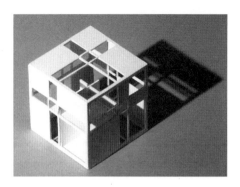

图 4.1-8
立方体练习示例；
Example of the cube exercise;

one of which is hierarchically dominant. All four spaces must be rectangular. The base for the walls is an 8 ½" X 11" rectangle also of cardboard." [fig. 4.1-7] These abstract exercises were designed to be do-able in a short period of time (2 hours) and constrained to focus on a single issue, the "lesson of the day".

Perhaps the most challenging abstract design exercise in the beginning year was the Cube problem. [fig. 4.1-8] The basic form of the brief was straightforward: construct a 5" cube using paper (2mm Strathmore white museum board). Inside the cube fix three rectangular planes at right angles to each other (i.e. defining xy, yz and zx orthogonal planes). The exterior faces of the cube must be composed of rectangular planes with no plane extending over the full 5" x 5" face of the cube. The nature of this assignment required precision and a high level of craftsmanship in the cutting and joining of the planes. This aspect of the problem, developing model-building skills, was however, not the main goal. The principle objective in the exercise was the strengthening of spatial perception. Alignment of the edges of the planes on the surface of the cube with those of opposite surface planes, as well as the three rectangular interior planes, enabled the articulation of a wide range of clearly defined rectangular volumes. An important consideration was the establishment of a spatial hierarchy. The idea was to avoid a 'checkerboard' non-hierarchical space definition. This naturally led to an asymmetrical distribution of planes and corresponding spatial volumes.

The Cube problem became a right of passage in the first year design studio at Cornell. Werner Seligmann, who championed the problem (and probably created it), believed that it enhanced spatial sensitivity and improved a student's ability to invent and manipulate space in a real design context. The Cube problem challenged the designer to subdivide the cube and organize a set of spatial volumes *sectionally* within the boundaries of a cube. Furthermore, the role of the surface planes composing the elevations

尽管康奈尔是一所特别推崇勒·柯布西耶作品的学校（柯布的作品全集是最基本的先例手册），其他建筑师、不同历史时期的建筑也会被用来讨论特定的形式属性——空间序列、立面构成或建筑与场地之间的关系。对于建筑形式结构的讨论构成了一个无关历史、无关风格的连续体，一套可适用于任何时空的建筑的基本原则。

例如，使用诸如位于罗马的朱利亚别墅的历史先例与想象它的某些方面（如从别墅入口、穿越后端半圆形庭院以及朱利叶斯三世设计的下沉式喷泉庭院的叠套的空间序列）在当代的建筑设计中进行变形或重新创造并不矛盾。"理想别墅的数学"一文通过对比16世纪的帕拉第奥别墅与勒·柯布西耶的早期现代住宅，来展示两者结构网格在比例与配置上近乎完美的接近。受到科林·罗这一比较分析方法的影响，学生开始有意识地选取相关的先例，抽取其中本质的形式特征作为启发，或是更为普遍地将它用以解决设计课题中的特定问题。

在本科的三、四年级，李·祝辰开设有一门连续三个学期的"现代建筑历史与理论"课程。课程以早期现代建筑运动为起点，祝辰通过对赖特、密斯、柯布以及阿尔托的作品进行细致而博学的分析来阐释现代建筑的基本原则。祝辰曾在阿尔托的赫尔辛基工作室工作过，他对于这位建筑大师的认识结合了个人的经历与第一手体会。第二学期的课程则接续介绍第二代现代主义建筑，并将目光主要集中在路易斯·康以及柯布后期的作品上。第三学期的课程则主要是通过分析意大利、法国的规则式花园来讨论建筑中的环境设计。祝辰同时还在四年级的建筑设计课程中，要求在一个开放场地上设计一个规则式花园。[图4.1-9]需新设计的人造花园所修饰的建筑选自现存的历史花园别墅，其布置应考虑建筑之于独特景观特征、花园模式的关系。如果现在开设一门类似的景观设计课程，其重点无疑是各种环境主题，包括绿化、水体保持与循环、私人庭院等。祝辰对课程的内容加以限制，把设计的重心放在诸如流线、序列、图形与几何、空间限定与剖碎等形式问题上。

20世纪60年代后期及70年代的康奈尔建筑学教学发展并改善了一种由柯布作品所派生出的形式主义设计方法。这一方法的基础则是与"自由平面"概念相

of the cube played a major role in the definition of the interior spatial volumes, reinforcing the modernist conception of an elevation that allows you to *read* the interior spaces of the building.

As students progressed through the studio sequence the role of precedent was continually revisited and applied on many levels. An obvious use of precedent is the notion of "building type", that is, associated with program or use. Precedents could also refer to formal typologies such as a courtyard precedent in housing. It is important to note that although Cornell was a school that venerated the works of Le Corbusier (the Oeuvre Complete was the essential precedent guide) many other architects and historical periods of architecture were referenced for specific formal attributes that might inform a decision about spatial sequence, the composition of a façade, or a building-site relationship. The formal structure of architecture was seen as belonging to an ahistorical, non-stylistic continuum; a set of fundamental principles applicable to architecture of any time or place.

For example, it was not contradictory to cite an historical precedent such as the Villa Guilia in Rome and imagine some aspect of it (e.g., the remarkable telescoping spatial sequence from the Villa's entrance, through the semi-circular courtyard in the rear, and its resolution in the sunken nymphaeum of the Pope Julius III) transformed and re-reinterpreted in a contemporary design proposal. "The Mathematics of the Ideal Villa" had after all compared a 16th century Palladian Villa with an early modern villa by Le Corbusier, revealing a near perfect match of proportion and configuration of the structural grid. Influenced by Rowe's comparative analysis technique, students consciously appropriated relevant historical examples, extracting their essential formal characteristics for inspiration or, more often, to resolve particular design issues in their studio projects.

At the third and fourth year level a three-term sequence in the History and Theory of Modern Architecture was offered by Lee Hodgden. Beginning with the early modern movement, Hodgden explicated the essential tenets of Modern architecture through detailed and erudite analyses of the major works of Wright, Mies, Le Corbusier and Aalto. Having worked in the Helsinki office of the latter, Hodgden offered a personal and firsthand knowledge of this master architect. The second course continued with the work of the second generation of Modern architecture, focusing primarily on Kahn and late Corbu. A third course was added to investigate *entourage* in architecture, mainly through the study of Italian and French formal gardens. Hodgden in parallel taught a fourth year architecture design studio with a project to design a formal garden on an open site. [fig. 4.1-9] The building embellished by the new contrived garden was selected from existing

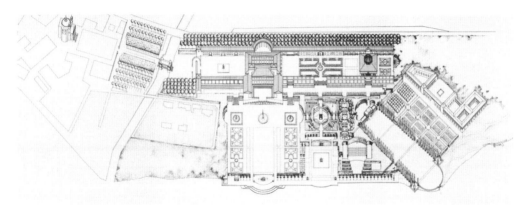

图 4.1-9
李·祝辰的规则式花园设计
课程作业示例，
作者：林伟而；
Example of Lee Hodgden's
studio on formal gardens,
design work by William Lim;

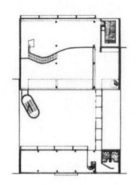

图 4.1-10
康奈尔学生习作示例，
其中强调自由平面及现代建
筑的品质；
Example of Cornell student
project with developed free
plan/modern architecture
qualities;

关的空间敏感。在实践中，它暗示了一系列策略的使用，包括空间区域与层次，均一而中性的结构网格（最为常见的是底层架空圆柱柱网），以及从剖面上利用重复、平直的水平楼板与屋顶板来发展空间。[图4.1-10]这一时期的学生作业表现出在特定场地、结构及功能限制下，空间组织与操作的精妙。设计过程的第一步是建立组织策略构思，这种有关模式的概念是布扎建筑设计方法的基础。通常来说，组织策略构思是一个平面图解，有时也可以是剖面图解。对于策略构思的找寻并不是随意的，它受到空间规划、功能使用计划分析、场地研究等多方面影响。在这一过程中，常常会有一个或多个建筑先例出现，它们与正在处理的设计问题就某一方面的特征或品质产生关联。当整体组织策略确定后，可以从一个暗示结构网格的骨架性底图开始，在一定比例尺的平面及剖面图纸上发展设计。

historic garden villas and sited according to the particular features of the landscape and the scheme of the garden. A similar studio on landscape design if taught today would no doubt focus on environmental themes; greening, water retention and recycling, private gardening, etc. Here the agenda was constrained to limit and thus focus design on formal issues such as circulation and sequence, pattern and geometry, spatial definition and poché.

The architecture program at Cornell during the late 60's and early 70's developed and refined a formal approach to design that was derivative from the work of Le Corbusier. Fundamental to this approach was a spatial sensibility that derived from the *free plan*. In practice, this implied the use of spatial zoning and layering, a uniform and neutral structural grid (most commonly a grid of round *pilotis*), and a sectional development of space within repetitive, flat, horizontal floor and roof plates. [fig. 4.1-10] Student projects of this era generally displayed an exceptional facility in manipulating spatial organization within site, structure and program constraints. The design process began with establishing a *parti*, the schematic concept that was so fundamental to the Beaux Arts methodology. The parti was most often a plan diagram, but could sometimes be based on the section. The search for a parti was not arbitrary and was initiated by space planning and program analysis, together with site studies. But invariably during the process one or more precedents would emerge, each containing perhaps a feature or quality that might have relevance to the project at hand. With a general organizational strategy set, the design could develop in scaled plan and section drawings, beginning with a skeletal base drawing indicating the structural grid. Tracing paper overlays provided the tool for transformation and spontaneous invention. Model studies in white Strathmore paper

对设计发展而言，不断叠加描图纸是一种有效的工作方法，它可以带来持续变化与自发创造。使用白色卡纸进行模型研究则有助于呈现三维的比例及空间关系。考虑到明暗、光影品质可强化对于边界与空间限定要素的感知，训练会优先使用白色的建筑模型（主要为纸模型）。

罗伯特·文丘里于1966年出版的《建筑的复杂性与矛盾性》是建筑理论发展的一个分水岭。甚至可以认为它是现代建筑权威受到的第一个重大挑战，它引导了因反现代建筑语言与正统而出现的后现代路线。这样的立场鼓励了一种更加大众化，以使用者为导向的建筑方向，这在查尔斯·穆尔、罗伯特·斯特恩以及罗伯特·文丘里等主要拥护者的作品中得到进一步宣扬。这些建筑师通过加入更多的细节（通常是历史的、复制性的），远离现代极简主义来创造一种新的反现代风格。在设计中使用色彩及装饰，与诸如彼得·埃森曼、理查德·迈耶、查尔斯·格瓦德梅、沃纳·塞利格曼等当代建筑师[10]采用的光洁的、受柯布启发的白色表面形成了强烈的对比。这一外观上的对比以及围绕这些立场的对立争辩带来了更为通俗的标签：“白色”与“灰色”[11]。［图4.1-11］

这些新的后现代思想对于建筑教育的影响是直接且广泛的。20世纪60年代的政治变化使得建筑领域出现了一种低技的、环境主义及激进主义倾向。其后，向形式主义及建筑自主性的观念转变受到了欢迎。“纸板建筑”正是形容代表这一风格的抽象化、形式化的模型（及建筑作品）技法的一个术语，它逐渐成为美国建筑设计教育中的标准。学生会争相模仿先锋设计师的作品，“明星建筑师”的说法也由此出现。在一所学校，究竟是“灰色”还是“白色”更占上风完全取决于该校的哲学立场。在康奈尔，渗透入教学的

were important in visualizing proportions and spatial relationships in 3D. White architectural models (primarily paper) were preferred for the sharpness of shade and shadow that enhanced the perception of edges and space defining elements.

The publication of *Complexity and Contradiction in Architecture* by Robert Venturi in 1966 was a watershed moment in architectural theory. It possibly marked the first serious challenge to the authority of Modern Architecture (MA) and led to numerous *postmodern* approaches that searched for alternatives to the language and orthodoxy of MA. One such position encouraged a more populist and user-oriented approach, as promoted in the works of its chief advocates: Charles Moore, Robert Stern and Robert Venturi. These architects created a new counter-modern style that incorporated more detail (often historical and highly derivative) and an intentional move away from the minimalism of MA. The introduction of color and ornament in their work, contrasted with the sleek, Corbusian inspired white finish of works by contemporary architects such as Peter Eisenman, Richard Meier, Charles Gwathmey and Werner Seligmann[10]. This contrast in appearance as well as the opposing polemics surrounding these positions, resulted in the popular labels of "White" versus "Gray"[11]. [fig. 4.1-11]

The impact on education of these new post mod initiatives was immediate and widespread. After a decade of political turmoil in the 60's that sponsored a low-tech, environmentalist and activist trend in architecture, the shift to a formalist and autonomous approach was welcomed. "Cardboard architecture", a term that typified the abstract, formal modeling technique (as well as the finished built work) of this style became standard studio practice in architecture education in the USA. Students emulated and imitated the work of the leading designers and the notion of "Star Architect" emerged. Depending on the school's philosophical position, either the "Grays" or the "Whites" attained prominence. At Cornell the position of Robert Venturi and the greys was antithetical to the Corbusian and more rational formal tendencies that pervaded the program. Aphorisms

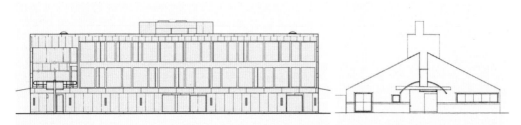

图 4.1-11

威拉德州立医院行政楼，沃纳·塞利格曼与母亲之家，罗伯特·文丘里；
Administrative Building at
Willard State Hospital,
Werner Seligmann and
Casa Vanna Venturi,
Robert Venturi;

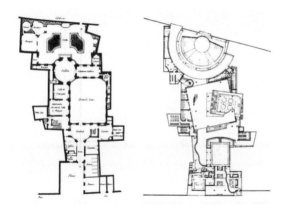

图 4.1-12

迈克尔·丹尼斯设计课习作：
维克弗朗什酒店与
约翰·麦克唐纳的法国威尼
斯研究所设计；
Example of Michael Dennis's
studio: Hotel de Villefranche
and John McDonald's design
of Venetian Institute in Paris;

图 4.1-13

康奈尔后期的学生作业中同
时包含图形化的及自由平面
的要素；
Project by Marianne Kwok at
Cornell with both figural and
free plan elements;

柯布派及更为理性的形式倾向与罗伯特·文丘里以及灰色派的立场是相对的。诸如"兼容并蓄"、"少是乏味"的标语被认为是为了批判主流现代建筑而吸引眼球的措辞与口号，它们与康奈尔学派针对现代主义所作出的修正是难以共处的。

20世纪70年代中叶，在迈克尔·丹尼斯的设计课中复兴了"剖碎"这一概念。[图4.1-12]丹尼斯此前对于"剖碎"的发展历程进行了系统研究——追溯它从布扎时期（16、17世纪法国酒店平面中的空间限定要素）到现代建筑（空间占据要素）的演进[12]。尤其是在勒·柯布西耶创造的自由平面中，将"剖碎"视为限定清楚的规则形房间之间的中介填充物的观点，在很大程度上已经消失。不具备用以储藏及设备装配所需要深度的薄分隔墙替代了有厚度、通常可容纳居住、经设计后具有一定灵活性的剖碎。空间中相应出现的是一些可感知为具体形式的独立支承物件，这些实体要素容纳先前剖碎的功能需求。

对于剖碎的重新认识并不局限于康奈尔大学。几乎是在同一时期，普林斯顿大学的迈克尔·格雷夫斯完成了从白色的柯布形式主义向关注传统房间及图形空间的后现代设计师的转变。虽然关于他的讨论常常与建筑的外立面处理有关（大概格雷夫斯最出名的是在外立面引入装饰性的历史拼贴与大胆的颜色），他对于剖碎以及图形空间构成的回归对传统的、前现代的空间规划的复兴有一定的贡献。在康奈尔，自20世纪70年代晚期起这样两种对待空间设计所相对的态度（自由平面与传统房间平面）相共存，甚至在有些时

such as "both-and" and "less is a bore" were viewed as catchy phrases, slogans critical of mainstream modern architecture and not compatible with the refined and edited approach to modernism of the Cornell school.

During the middle of the seventies decade, a re-immergence of the concept of *poché* appeared in studios taught by Michael Dennis. [fig. 4.1-12] Dennis had spent a few years researching the evolution of poché from its beaux-arts manifestation as a *space definer* in the plans of 16th and 17th century French hotels thru its transformation in modern architecture to that of a *space occupier*[12]. Especially in the free plan as envisioned by Le Corbusier, poché as in-between filler shaping the regular and defined space of the room, had largely disappeared. Thin partition wall dividers lacking the depth needed to provide for storage and the accommodation of utilities replaced thick, deep, often habitable, and by design, flexible poché. In its place the freestanding object, an elemental solid strategically located in space to be perceived as form while at the same time accommodating the former functional provisions of poché was born.

The re-discovery of poché was not restricted to Cornell. At about the same time, Michael Graves at Princeton completed his transformation from "white Corbu formalism" to that of a postmodernist designer of traditional rooms and figural space. Although his reputation was mainly associated with the exterior treatment of the building (Graves is probably best known for introducing decorative historical pastiche and bold color rendering of the surfaces) his return to poché and figural space composition contributed to a revival of traditional, pre-modern space planning. Beginning in the late 70's at Cornell, these two opposite approaches to spatial design, the free plan and the traditional room plan, coexisted and sometimes merged together to produce a hybrid modernism that was truly "both-and". [fig. 4.1-13]

候相互融合以产生一种混杂的现代主义，这才是真正的"兼容并蓄"。[图4.1-13]

第三节：教案 | 俄亥俄州立大学 1985-1991

1983至1985年间，俄亥俄州立大学的建筑学院来了一批新的年轻教员，其中突出的有罗伯特·利夫西，他是普林斯顿的毕业生，曾在格雷夫斯门下学习。俄亥俄州立大学的建筑教育有着悠久的历史，其教学讲求实用性，以专业实践为导向[13]。利夫西于1983年成为建筑系的主席，在他的带领下该学院向国际化方向扩展，并发展出一条具有批判性、重视理论的教学路线。

1985年，建筑学院开设有四年制理学学士的职业预备学位课程以及两年制的建筑学硕士学位课程。作为一所公共赠地的大学机构，俄亥俄州立大学对于本科课程有一个公开的招生程序。任何希望进入该大学工程学院的学生可以在第一年选择建筑学作为他们的专业。由于这么专业特别受欢迎，申请的学生数量远远超过了学院条件所允许的数量，这时候就需要1个用以筛选的平等竞争制度。解决的办法是在一年级开设一门入门设计工作室课程，以此来决定学生是否具有在该专业继续学习的能力。

我当时被任命为这门一年级入门建筑设计工作室的主持人，我将这门课程分为一个为期10周的模块系列[14]，其中包括6个为期1周的设计练习以及1个为期4周的复杂设计问题。选修这门课程的学生大概有120人，每周在配有大工作台的教室里安排2次长3小时的面授。为了应对如此大规模的学生，我招募了8到10名建筑系的研究生作为助理导师。每一位助手需要负责指导工作室内的12至15名学生。每周我会有一次为全班同学准备的讲座，助理导师也会参加。另外，我每周还会额外约见这些助手一次，说明接下来要进行的练习并讨论内部的一些事务。有时候这种约见可以看作是非正式的"教室培训"课程，涉及的内容包括工作室教学技巧、"每日一题"的特定教学目标等。

这一入门设计工作室采用的是非指定的临时座位制度，学生每次上课都要带上他们自己的工具[15]。因为受到工作环境的时间、空间等限制，每一次课都需要精心设计：限定范畴、易于执行以及主题明确。其中的练习最初来自于康奈尔开设的一门同类型课程。一

Part 3: The Program | Ohio State University 1985-1991

Between 1983 and 1985 the School of Architecture at the Ohio State University (OSU) received an influx of young new faculty led by a Princeton graduate and student of Michael Graves, Robert S. Livesey. Architecture at OSU had a long tradition and the program was regarded as pragmatic and practice-oriented.[13] Under Livesey, who became Chair of the department of Architecture in 1983, the school expanded its international scope and developed a more critical and theoretical approach.

In 1985 the School of Architecture had a pre-professional four-year Bachelor of Science degree program and a two-year Master of Architecture. As a Public Land Grant institution, OSU had an open admissions process to undergraduate programs. Any student entering the College of Engineering at OSU could elect to study architecture as a subject in the first year. As a result of the popularity of the major, the number of applicants to the program far exceeded the number of places available and a competitive and fair system for placement was needed. The solution was the creation of an open introductory architectural design studio in the first year that would determine the eligibility of the student to continue on in the major.

Appointed coordinator of the first year introductory design studio, I structured the course as a 10-week series of modules[14] consisting of six 1-week design exercises and a longer, 4-week comprehensive design problem. The enrollment for the course was approximately 120 students, meeting twice a week for three hours each session in a classroom equipped with large flat working tables. To manage the numbers, I recruited 8-10 graduate students in architecture to serve as assistant instructors. Each assistant was responsible for guiding 12-15 students in the studio session. Once a week I conducted a lecture for the entire class with the assistant instructors attending. In addition, I met separately once a week with the assistants to go over the exercise for the next session and to discuss any

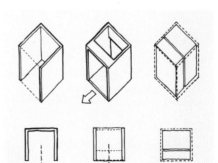

图 4.1-14
6片板的立方体限定练习的设计研究；
Design study of Definition of a cube with six planes;

些短周期的练习（包括"25片板的空间限定"与"6片板的立方体限定"［图4.1-14］）旨在训练学生在一定的限制下快速掌握空间形式的可视与操作。

这门课程的重心在于通过一系列的抽象形式设计练习来训练空间的设计及构成。这一做法将对于功能、文脉、建筑技术等其他设计问题的介绍延后，以此降低这一阶段设计工作的复杂性。对于教授设计入门的老师而言，是否在一开始就引入关于建筑设计的全部问题是关键性的决策之一。一些教师更倾向于一种包容主义的方式，他们认为在基础设计问题中排出这些问题会规避与现实情况（如真实的场地）的必要接触，这样的做法会导致一个简单化、主观的形式抽象[16]。

我们的经验是，学生是需要分阶段吸收设计知识（如理解真实场地的物质属性）并逐步发展出一个综合而全面的设计方法。在人文与技术等支撑性课程中获取的知识可以提供必要的基础，而使得学生能够完全掌握建筑设计的各细节及其复杂性。我们所倾向的这种方式强调空间在建筑中的核心角色，减少（而并非是完全消除）限制以集中关注于视觉及空间素养的培养。

设计中主观性的假设所暗示的是设计价值或品质的不定性，每个人对此的解读与评价会有所差异。这里需要争辩的是，如果跳脱出传统的评价标准（诸如对场地的回应，环保性能，经济效益，对于功能的组织等），对于一个抽象空间与形式构成的评价可集中于它的视觉特征，包括清晰的形式组织，独立要素的比例与层级关系，以及更为重要的空间限定的可读性[17]。在这种情况下，作品的内容是构成，一个设计的成败或品质在一定程度上取决于该构成与设计概念之间的连贯，这些是通过形式来表现的。

为了增加初学者对于设计问题多样性以及建筑文脉现实的认知，这门课程使用了大量的建筑先例。具体的做法有两点：其一，在讲座中对于案例进行介绍与评价；其二，在课程阅读材料中加入建筑先例及其图解分析[18]［图4.1-15］。讲座中讨论的先例主要是用来解释并强化对于形式原则的理解。与此同时，会通过讲解与该案例相关的设计问题，以提供一个比较全面的认识。举个例子，为说明赖特所设计的马丁住宅（1906年）中开放而复杂的空间层次，就需要同时介

internal issues. Many of these sessions also served as informal "teacher training" classes with topics ranging from studio teaching technique to specific objectives of the "lesson of the day".

The beginning design studio used "hot desks" (temporary and unassigned seating) and students brought their equipment[15] to studio each class period. With time and physical constraints on the working environment, lessons were devised to be limited in scope, simple in execution and thematically focused. Initially the exercises were adopted from similar courses I was familiar with at Cornell. Short exercises such as *space definition with 25 planes* and *definition of a cube with six planes* [fig. 4.1-14] enabled students to quickly visualize and manipulate space form with limited means.

The primary focus of the course was space and composition and all of the exercises were designed to address these two themes. Abstract formal design exercises executed in drawings and models formed the basis of the course work. Such an approach postpones the introduction of function, context, building technology and other design issues, in part to reduce the complexity of design at this stage. This is one of the crucial decisions in beginning design education; whether or not to engage from the outset the full range of issues that inform architectural design. Some educators favor an *inclusivist* approach arguing that eliminating these issues in a beginning design problem avoids the necessary confrontation with real conditions (e.g. an actual site) and leads to a simplistic and subjective formal abstraction[16].

Our experience is that students absorb design knowledge (e.g. understanding the physical attributes of a real site) in stages and gradually develop an integrated and more holistic design method as they progress through the design curriculum. Knowledge gained from supporting courses in humanities and technology provides the necessary foundation that eventually enables design students to fully grasp the detail and complexity of architectural design. The approach that we favor emphasizes the primary role of space in architecture and minimizes (but does not eliminate entirely) other parameters so as to focus on the development of a visual and spatial literacy.

The assumption of subjectivity in design, implies that the merit or quality of a design is mutable and open to the interpretation and judgment of the individual. We would argue, however, that even without the conventional methods of assessment, for example, the response to site, environmental performance, economy of means, accommodation of use, etc., an abstract composition of space and form may be evaluated purely on visual character, that is; a clarity of formal organization, the proportion and hierarchy of individual elements, and most importantly, the legibility

绍其中结构性砌体柱的排布方式、空间使用的功能计划、流线模式等。而阅读材料中则选取了诸多建筑及绘画、艺术作品，并通过图解的方式来反映该作品中的形式原则。另外还有包括一些与理论相关的短篇文献。

1. 设计工作室练习

对于大部分设计的初学者而言，缺乏基本的绘图技能是一个不争的事实。虽然有一些学生曾学习过使用尺规的技术制图（可称之为"硬笔"绘图），但是比较少的学生接受过草图或徒手画的训练，亦不了解它在设计学习中的关键作用[20]。这也是为什么这门课程在一开始需要介绍绘图的相关内容（即本书的第一部分"技巧"），通过大致3周的时间来学习徒手草图及建筑图学基础。我们使用了一些漆白的木块及圆柱体，将他们成组布置在一起以形成一个简单的、具有建构属性的物件组合，如静物一般。放置这些物件的底板上划分有网格，它与其他的物件在尺寸上共用一套模数系统。在这种情况下，使用网格纸来练习制图可以简化学生徒手绘制各平面图纸的难度。物件本身的模数关系使得它们可以通过整数比例来表示其各边尺寸（如1x2x4），这样就可在不诉诸于特定比例尺的情况下进行准确的绘制。这样的做法是有目的性的，它可以为学生建立起对于比例的概念，加强徒手制图的准确性。在整个课程中，每一个练习都有对应的绘

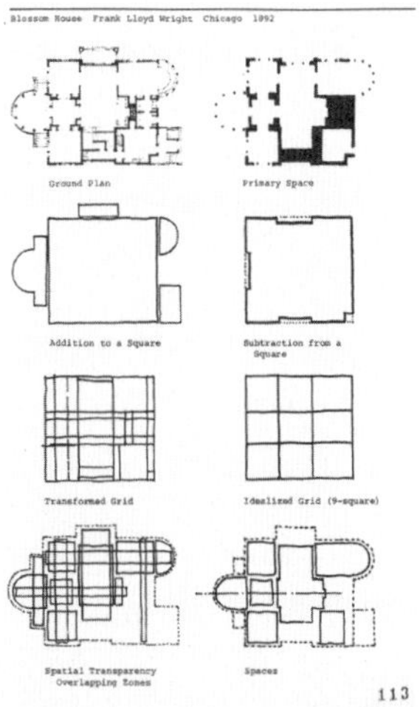

图 4.1-15

布劳萨姆住宅图解分析，《传统平面与自由平面》内页；
Diagrammatic study of Blossom House, page extracted from *Plan Trad* | *Plan Libre*;

113

of spatial definition[17]. In this sense the *content* of the work is the composition, and the success or quality of the design resides to some extent on how effectively the composition communicates the design concept as expressed in the form.

To provide beginning students with a broader exposure to the multiplicity of design issues and contextual reality of architecture, the course makes reference to a wide spectrum of architectural precedent. It does this in two ways: first, through the presentation and critique of examples in the lecture class, and second, by the inclusion of precedents and their diagrammatic analysis in a course *reader*[18] [fig. 4.1-15]. Precedents discussed in lecture are selected to explain and reinforce an understanding of formal principles. At the same time reference is made to relevant issues critical to a broader understanding of the work. For example, an appreciation of Wright's open and complex spatial layering in the Martin House (1906) would be incomplete without reference to the logic and controlled placement of the structural masonry piers, the program of space usage, the circulation patterns, and so forth. The reader included select examples of architecture and art (paintings) with diagrams that revealed the formal principles within the work. In addition, selected short essays pertaining to theory were also included[19].

1. The studio exercises

It is a fact that most beginning design students lack basic drawing skills. While some students may have been exposed to technical drawing with instruments (referred to as "hard line" drawing), few are guided in the technique of sketching or "freehand" drawing, and its critical role in learning[20]. By necessity the course introduced drawing at the outset and the first section referred to as "Technique", devotes about three weeks to learning the basics of freehand sketching and architectural drawing convention. Small model setups consisting of painted wood blocks and cylinders arranged in a group form simulates a simple, architectonic construct as a still life. The base for the objects has a grid inscribed on it with a module compatible with the dimensions of the block elements. Using a gridded sheet of paper for the drawings simplified the freehand construction of planimetric views. The modularity of the blocks having whole number ratios for their dimensions (e.g. 1 x 2 x 4), enables accurate drawings to be constructed without resorting to a scale. This is intentional and helps the student develop a sense of proportion as well as precision in freehand drawing. Throughout the course, drawings for every design exercise were required. These included documentation in plan, elevation, section, and

图训练，它包括记录性的平、立、剖面图与轴测图，以及分析性的图解。训练允许学生使用尺规绘制一层轻轻的线稿作为底图，但是最终成图必须是描摹其上的徒手铅笔线条。

课程的第二部分"形式"介绍了空间及其相对的实体或体量概念。为研究实体与虚空的特征，一组相互配对的设计练习（"实体"与"空间"）应运而生。第一个练习使用一组共11件的模数要素来构成一个单一的实体物件，第二个练习则要求使用相同的部件要素来限定一个单一空间容积。两个练习都强调单一形式的创造，第一个是一个物件实体而第二个是一个通过限定而获得的具有尺寸及可读形式的空间容积。最终创造的形式应当是明确的。

设计这两个相互平行的练习是对于教学法及实践必要性的回应。从教学法的教学来说，将空间感知为形式与更本能地将物件感知为形式相矛盾，这一意识的转变对于一个初学者来说是关键性的一步。与此同时，在练习中又引入了"加法"与"减法"操作的概念。从课程执行的教学来说，练习应该限制在4个小时时间内（2个周期时长），易于模型操作，尺寸上满足于A3绘图纸上的等比例绘图。通过设计一套模数化的要素（较小的要素通过叠加在一起可获得较大要素的尺寸）作为装配部件是具体突破性意义的。一方面，它确保了大部分设计成果的美观、另一方面它使得加工这套装配部件中的各木质要素更为简单。

课程接下来两部分被称为是"构成"。控制视觉秩序的诸规则（轴线与对称，层级、基准等）被介绍并在简单而抽象的，以模型发展为手段的设计练习中得到应用。此外，还介绍有关于比例的概念，并探索在艺术与建筑这两个领域中的图底概念。"构成"这一部分最初与舍伍德的《建筑的原则与要素》紧密结合，设计教学也受到早期得州一康奈尔发展出的设计练习（如用25片板限定空间）的影响。每个练习都聚焦在一个或两个问题上，并确定一系列的参数来限制设计操作的可能。

将这一思路继续延续，练习基准墙所讨论的是以空白场地内一个既存的物件为起点，围绕它进行"文脉"设计。这一物件是由练习的第一步所准备的，它是一个给定尺寸的墙体要素，其上有部分体量被挖去以形成开口或壁龛。练习同样以一组装配部件为起

axonometric as well as analytical diagrams. A very light mechanically drawn underlay in pencil was permitted, but the finished drawing is a freehand pencil trace over the grid guideline construction.

The second section of the course, Form, introduced the concept of space and its opposite, solid or mass. Two complimentary exercises were created (Mass and Volume) in order to study the formal characteristics of solid and void. The first exercise employed a set of 11 modular elements to compose a singular solid object while the second used the same kit-of-parts to define a singular space or volume. Both exercises emphasized the creation of a singular form, the first an object mass and the second, a defined spatial volume with dimension and legible form, absent any ambiguity.

The invention of these two parallel studies was a response to both pedagogical and practical necessity. Pedagogically the perception of space as form was contrasted with the more intuitive and preconceived view of object as form. The awareness of solid-void reversibility is a critical step for the beginning student. At the same time notions of additive and subtractive process were embedded in the exercise. Administratively the exercise needed to be limited to about 4 hours (the length of two sessions), easy to manipulate in model form, and of a size that could be drawn at full scale on A3 size paper. The design of a modular set of elements (dimensions selected so that smaller elements add together and "match" the dimension of larger elements) as a kit-of-parts was something of a break through. Not only did it contribute to the elegance of many of the designs[21], but it also made the prefabrication of the wood pieces constituting the kit much simpler.

The third and last section of the course is called Composition. Here the rules that underlie visual order: axis and symmetry, hierarchy, datum, etc., are explained and then applied in simple, abstract design exercises developed in cardboard models. Proportion is also introduced and the concept of figure and ground is explored in relation to both art and architecture. This section was initially tied to Sherwood's *Principles and Elements of Architecture*, with exercises that evoked many of the early TU-Cornell problems such as *space definition with 25 planes*. Each exercise focused on one or two issues at most with a set of parameters determining the operational constraints.

Extending this approach, the exercise datum wall was created to introduce the formal inter-relationship between object and context. The idea was to work in reverse, that is, take a formed object, place it on an initially blank field and then create the context. The object, prepared as step one of the exercise, was a wall element of predetermined dimension, modified with a few openings and notched subtractions. The kit-of-parts was again used and included a range of object elements (cube, cylinder,

点，包括的物件要素与先前的练习相似，包括立方体、圆柱体、矩形棱柱以及墙体板片。设计的目的是在整个方形场地上创造限定空间区域的构成。一个相对成功的设计往往会将空间组织与基准墙、方形场地相结合，使用加法、减法的操作方式，并结合五种秩序原则来达到一个整体并清楚可读的形式设计概念。

2. 重新创造的九宫格问题

大致是在20世纪70年代中期，我开始注意到位于纽约的库伯联盟建筑学院。在拜访这所学校的时候，我注意到这里新装修的优雅的白色室内，同时更吸引人的是分布在这座楼里面的学生作业。工作室里满是精细的木模型，细致的硫酸纸墨线图纸，以及伴随其侧的徒手铅笔草图、水彩颜料与印刷品交叠的复杂拼贴。其中有一个房间让我印象特别深刻，在绘图桌上摆放着一个个帅气的木盒子，里面整齐的包装着很多形状各异的实体物件，包括正方体、圆柱、平板等以及一个底板，其上有由杆件建构而成的方形三维网格，形同梁柱框架系统。[图4.1-16]这一框架是方形的，它由9个均分的单元模块组成。空置的白色框架，除去梁柱之外并不包含有任何其他要素，投射出对一种强有力的建构的纯粹性，并暗示一种身处建筑建造中的状态。

这套框架以及装配部件是约翰·海杜克的发明，它可以看作是建筑本质的理性抽象。库伯联盟一年级的学生会使用这个框架作为他们第一个建筑设计的"场地"。它是一个具有启发性的媒介，以此可以介绍空间的构成与形式设计的语言。在框架中布置选定的

and rectangular prisms) plus the wall planes as in the earlier exercise. The design objective was to create a composition of defined spaces and spatial fields over the entirety of the given empty square site. A successful design would create a pattern of spaces related to the datum wall and the square field, would use additive and subtractive design process techniques, and employ ordering principles to achieve unity and a clear and legible formal design concept.

2. The nine-square problem re-interpreted

At some point in the mid-seventies I became aware of the Cooper Union School of Architecture in New York City. Upon visiting the school I was struck by the white elegance of the recently renovated interior, and most of all, by the proliferation of student work throughout the building. Stunning wood models, high precision drawings of ink on Mylar™ side by side with evocative freehand drawings in soft pencil and complex collages of print media overlaid with gouache paint filled the studios. But in one room I came across something even more fascinating. On the drawing tables were handmade wood boxes containing a range of different shaped solid objects neatly packed away: cubes, cylinders, flat planes, etc., and a base with a three-dimensional grid of square sticks forming a frame of columns and beams. [fig. 4.1-16] The geometry of this frame was square and it consisted of nine equal bays or modules. The emptiness of the white frame, the absence of anything other than the columns and beams, projected a powerful sense of architectonic purity, and the suggestion of an architectural construct in progress.

The frame and kit-of-parts was the invention of John Hejduk and it was a cerebral abstraction of the essence of architecture. Students in the first year studio at Cooper Union used the frame as the "site" of their first archi-

图 4.1-16
库伯联盟九宫格练习的装配部件；
The kit-of-parts of Nine-square problem at Cooper Union;

图 4.1-17
网格框架（5x5）练习习作；
Example of Grid Frame (5x5);

物件可以直接引出关于视觉的对话，可以通过持续的模型操作来测试不同的设计可能与变形。

对于我在俄亥俄州立大学教授的课程而言，我需要具有一定复杂性的设计课题，它能够延续数周时间，并在这一练习中融汇前面所有短练习所研究的概念与原则，以此作为课程训练的一个顶点。我还记得库伯联盟所使用的网格框架，然后决定发展一个类似的练习：在同样尺寸的方形底板上布置5x5的方格网。我设想通过增加的方形模块来延伸场地，可以增强构成的丰富性，也有可能创造一个更加显著的图底关系。[图4.1-17] 另外，我意识到实体物件的密度不够，这导致诸多设计方案都只是按照一定图示来进行简单的物件布置。

因此，我决定通过使用有别于框架结构的另一系统来改变关于密度的问题。厚墙似乎是一个可能的选择。在场地上布置有一组厚高比为1:6的平行墙体，减少它们彼此间的间距以达到约20%的实体面积。练习中可以重新调整墙体的位置、缩短其长度或是在上面开挖洞口以创造通道。在墙体之间增加横跨墙体的水平屋顶板片，这样便完成了假设的承重结构。在这个练习中，同样提供了一套具有模数关系的部件要素以及高度低于平行墙体的薄的墙体板片。这些要素可以与厚墙体产生对话，正如基准墙练习一样。这样，一个全新的平行墙练习便开始了。

这一设计练习经历过多次调整，但是一直保持了平行墙这一基本体系以及对额外部件要素的使用。正如设计目标所言——"入口位于与墙体结构相平行的场地边界上，并以此展开游览的流线及空间序列"，这个练习是充分开放与抽象的，它可以充分调动学生的创造力，亦可对设计的功能或目的展开想象。这是一个有院子的住宅，还是类似于金博尔的一个博物馆？又或者是受阿尔多·凡·艾克早期作品启发的一个儿童游乐场？学生在练习中投射特定使用功能计划的做法并不会受到反对，它与练习本身的形式挑战亦不会相互干扰。

这一设计练习的成果关键在于设计任务书中强加的限制条件，这些是需要严格遵守的。这些条件控制了平行墙体的布置与开启，屋顶要素的摆放与形状以及对于不受限使用的分隔墙体的特定要求。这些条件减少了学生在设计过程中的分心，让他们集中注意在

tectural project. It was a heuristic device to introduce spatial composition and the language of formal design. Placement of selected objects within the frame immediately elicited a visual dialogue, one that could be manipulated and tested through the variations and transformations of subsequent modeling.

For the course I was teaching at Ohio State I needed a project of some complexity, an exercise that would extend over several weeks and serve as a kind of capstone design project incorporating all of the concepts and principles explored in the preceding short exercises. I remembered the grid frame from the Cooper Union and decided a similar exercise employing a 5x5 square grid frame on a square base of the same dimension might provide some challenge. I imagined that the extended field of additional square modules would allow for more compositional richness and perhaps a more pronounced figure-ground relationship. [fig. 4.1-17] Instead, after several iterations of this exercise, I found that the density of the solids was too small and that many of the projects could not escape from the easy solution of arranging objects in a symmetric pattern within the frame. The symmetry and regularity of the grid frame was too strong to overcome.

In reaction I decided that the problem of density and the regularity of the grid might be resolved by adopting an alternative to the frame structure. Thick walls seemed to offer a solution. A field of parallel walls of a proportion of 1 to 7 (thickness to height) with relative close spacing introduced a solid fill of about 15%. Critically, the walls could be repositioned (parallel to the edges of the base and each other but varying in spacing), shortened in length, and openings introduced to imply passage. The addition of roof planes spanning between the walls completed the assumed load bearing structure. As in the grid frame, a kit of modular object elements including an unlimited number of thin walls of a lower height than the parallel walls (and free to be positioned in any direction) was provided to be introduced into the design, potentially in dialogue with the thick walls, as in the datum wall exercise. So began the *parallel wall* exercise[22].

Through the many iterations of this problem the basic system of the parallel walls and the additive elements of the kit-of-parts remained a constant. The stated objective of the design, "create a circulation and spatial sequence initiated from an entrance on one of the boundary edges parallel to the wall structure" was sufficiently open-ended and abstract to allow a large amount of formal creativity as well as imaginative speculation on the purpose or function of the design. Is it a house with courtyards? A museum recalling the Kimball? Or perhaps a children's playground evoking an early project by Aldo Van Eyck? The projection by the student of a particular func-

211

空间限定及序列的训练上。在清楚意识到设计目标的同时，在给定的限制范围中进行设计，这是具有挑战性的，当然也是一种十分有效的训练方式。对于空间规划的要求，对于结构可能性的考虑，对于功能与使用的关注是一位设计师在找到最终方案之前所必须跨越的限制。

3. 超越平行墙练习的抽象性

　　乔尔乌及萨拉巴伊别墅使用的都是承重平行砌体墙构造，这一类具有明确教学价值的建筑暗示了平行墙练习可以在更为传统的高年级设计工作室课程中作为一种设计课题载体。这样的形式设计如何回应功能计划？采用一个真实（或想象出来）的场地？怎样的材料与建造体系最为合适？

　　这一系列疑问在之后的二年级设计课程中得到了解答，平行墙练习作为一个入门性的研究课题[图4.1-18]安排在了长周期的设计课题（艺术工作室社区大楼）之前。设计任务书明确了功能计划的必要部分，包括独立的工作室生活单元、工坊、餐厅、室外空间等。这个场地是虚构的：近水且周围无其他建筑物，位处亚热带气候条件。最后，该建筑设计的构造系统为平行混凝土砌体墙，其上支撑预制的屋顶模块。这一设置的目的是发掘空间设计的形式可能以及抽象研究中形式秩序，然后在之后更为实际的建筑设计问题中予以应用。

　　事后想来，当时在库伯联盟的原始九宫格问题从未忽略建造问题，而是通过一个清楚的、教谕式的举动来介绍结构与空间之间的逻辑与相互作用，这样对于设计初学者而言可能是可以理解的。正如海杜克在展览手册中针对九宫格问题所言："通过这个训练，学生开始探索并理解建筑的要素——网格、框架、柱、梁、板、中心、外围、场地、边缘、线、面、体量、伸展、压缩、张拉、剪切等。"[23]

tional use or program on the evolving design was not discouraged nor did it interfere with the formal challenge of the exercise.

A key to the success of the problem is the *limitations* imposed in the brief or handout. Strict adherence to the rules that govern the rearrangement and opening up of the parallel walls, the positioning and shape of roof elements, and the conditions placed on the use of partition walls, all work to minimize distractions and strengthen the focus of the exercise on space definition and sequence. The struggle to realize a design intention while working within the parameters of the problem is a valuable lesson in itself. In actual building design, space planning, structure, function and use, all impose restrictions that a designer must negotiate to achieve a cohesive design.

3. Beyond the abstraction of the parallel wall exercise

The didactic clarity of buildings such as the Villas Jaoul and Sarabhai, both parallel wall load-bearing masonry construction, suggested the parallel wall exercise as a device for a more traditional studio project in a higher level studio. How would the formal design respond to a program? A real (or imagined) site? And what materials and construction system would be appropriate?

These questions were addressed in a year-two studio where the parallel wall exercise was introduced as a research study [fig. 4.1-18] prior to work on the longer term design project, an art studio community building. The brief identified the required components of the program: individual studio live-in units, workshops, canteen, outdoor spaces, etc. The site was an imaginary plot near water with no built context. The climate was semi-tropical. And last, the construction system of the building would be parallel concrete masonry walls supporting pre-fabricated roof modules. The intention was to discover the formal potential for spatial design and order in the abstract study and then carry the lessons learned into a more realistic architectural problem.

In hindsight the original Nine-square problem at the Cooper Union was never intended to ignore the issue of construction, but to introduce the logic and interplay between structure and space in a clear, didactic manner that would be comprehensible to the beginning design student. As John Hejduk stated in the forward to the Nine-square problem published in the MOMA catalogue of the show: "working with this problem the student begins to discover and understand the elements of architecture. Grid, frame, post, beam, panel, center, periphery, field, edge, line, plane, volume, extension, compression, tension, shear, etc."[23]

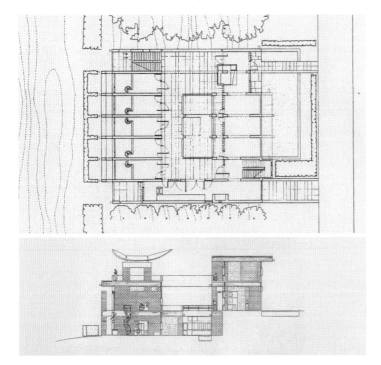

图 4.1-18

以平行墙练习为原型的二年级建筑设计课题;
Second-year design project
based on Parallel Wall project;

注解 notes

1. The first edition of *Architecture: Form, Space and Order* (Francis D.K. Ching) was published in 1979 and quickly became a classic reference manual for students of architecture. The presentation of "point-line-plan-volume" in the beginning of the book, along with the section on "Principles", draws heavily on the publication of the class notes of Roger Sherwood entitled *Elements and Principles of Architecture*.

程大锦所著的《建筑：形式，空间与秩序》（第一版）于1979年出版，并迅速成为建筑学学生的一本经典参考书。该书开篇对于"点-线-面-体"的陈述以及其后"原则"这一章节都深受罗杰·舍伍德的课程讲义《建筑的要素与原则》的影响。

2. It was at Cornell University around 1962 that the name "Texas Rangers" was initially used to identify those teachers (John Shaw, Lee Hodgden and Werner Seligmann) who had recently come to Cornell from the University of Texas. The origin of the title seems to have been taken from a parody song written by either Alan Chimacoff or Thomas Schumacher, students at the time. Alluding to the infamous law enforcers in the early days of the Wild West, a Texas Ranger had to "ride a horse, shoot straight and have a good eye" (i.e. possess the pre-requisite skills, stay true to the path of modern architecture, and be visually perceptive).

在1962年左右的康奈尔大学，"得州骑警"一名号最初被用来指代刚从得州大学过来康奈尔的一批教授（约翰·肖，李·祝辰以及沃纳·塞利格曼）。这一名号最初似乎是来自于当时的学生艾伦·奇马科夫与托马斯·舒马赫所写的一首改编歌。"得州骑警"暗指之前"狂野西部"年代声名狼藉的执法人员。骑警需骑马、准确

射击与好的视力，换句话说就是掌握必须的技能，忠实于现代建筑，对视觉的领悟力强。

3. Alexander Caragonne. *The Texas Ranger: Notes from an Architectural Underground*. Cambridge, MA: The MIT Press, 1995: 33.

4. Ibid.

5. "Transparency: Literal and Phenomenal" was the title of two essays written by Robert Slutzky and Colin Rowe in the mid-fifties. "Transparency I" was first published in Perspecta 8 in 1964 while "Transparency II" in Perspecta 13/14 in 1971.

"透明性：物理层面与现象层面"是罗伯特·斯洛茨纪与科林·罗于1950年代中期所著的两篇论文的题目。"透明性1"最初发表于1964年出版的《眺望》杂志第8期，"透明性2"则发表于同一杂志的1971年第13/14期。

6. *Notes on Saper vedere l'architettura*. Bruno Zevi (1948). Translated into English as: *Architecture as Space: How to look at Architecture* (NY: Horizon Press, 1957).

布鲁诺·赛维于1948年出版《建筑空间论：如何品评建筑》，该文原文为意大利语，于1957年出版英译本。

7. John Hejduk. *John Hejduk, 7 Houses: January 22 to February 16, 1980*, Catalogue - Institute for Architecture and Urban Studies; 12. NY: Institute for Architecture and Urban Studies, 1979.

8. Colin Rowe. *The Mathematics of the Ideal Villa, and Other Essays*. Cambridge, Mass: MIT Press, 1976.

9. The name Leporello came from the name of the character in Mozart's opera, *Don Giovanni*, who in Act II unfolds a long folded paper that contains a list of names of the many women seduced by the Don.

莱波雷洛来自于莫扎特创作歌剧《唐·乔望尼》（唐

璜）中的一章节，他在第二幕中展开了一张折叠过的纸，其上列有被唐所吸引的女子的名字。

10. Werner Seligmann had an outstanding small practice in upstate New York, winning numerous design awards. His buildings exemplified a refined modernism derivative of Le Corbusier. Interestingly, there was also a marked Frank Lloyd Wright influence in his earlier work.

沃纳·塞利格曼在纽约北部曾有一个出众的小型建筑实践，该项目有获得很多奖项。他的建筑作品展示了一种源起于勒·柯布西耶，后经精炼的现代主义。有趣的是在他早期的作品中同样能看到赖特的影响。

11. The "Whites" were canonized in the publication of *The New York Five* that contained a preface written by Colin Rowe. The five included Eisenmann, Meier, Charles Gwathmey, John Hejduk and Michael Graves (based on his early work such as the Hanselmann House). Werner Seligmann was not included primarily because of his practice being located outside of the New York City region. He was often later referred to as the sixth member of the Five.

"白色派"一词在《纽约五人》一书受到推崇，科林·罗为该书著前言。五人包括艾森曼、迈尔、查尔斯·格瓦德梅、约翰·海杜克与迈克尔·格雷夫斯（根据其早期作品，如汉索曼住宅）。沃纳·塞利格曼不在其中的主要原因是其实践在纽约市范围外，他常被称为"五人组"的第六人。

12. Michel Dennis. *Court & Garden: From the French Hôtel to the City of Modern Architecture*. Cambridge, Mass.: MIT Press, 1986.

13. The OSU School of Architecture began as the 11th architecture program in the United Sates in 1899. In 1994 the school became the Austin E. Knowlton School of

Architecture.

1899年出现的俄亥俄州立大学建筑学院是美国第11个开设建筑学专业课程的学校，该校于1994年改名为奥斯丁·诺尔顿建筑学院。

14. OSU at the time was on the trimester system that divided the academic year into three regular terms of 10 weeks in duration and a summer term also of 10 weeks.

当时的俄亥俄州立大学采用三学期制，即将一个学年分为三个长十周的学期，包括一个夏季学期。

15. A tool kit consisting of standard drafting equipment including a roll of tracing paper and a 30 inch drawing board was available in the bookstore at a discount for the set. A conscious decision was made not to introduce computational design drawing.

学生可以在校园书店里以折扣价购买一个由标准的绘图工具（一卷草图纸以及一块30英寸幅面的绘图板）组成的基本工具箱。这一做法的主要目的是排除电脑制图。

16. Robert Somol summed it up succinctly in his essay "Form en abyme" for the National Conference on the Beginning Design Student held at the School of Architecture at Louisiana State University in 1996: "The power and beauty of the nine-square problem, of course, was that it was immaterial, that it existed without function, site, client, body and, to some extent, even without scale. And this abstract universality was precisely, in the multicultural 1980's and 90's, what came to be challenged as exclusive, irrelevant, and closed; judgments which mapped too closely to the failure of the profession itself to go unheeded." Access: http://ncbds.la-ab.com/Batture_PastPapers.pdf

罗伯特·苏摩在"形式嵌套"一文中总结："九宫格问题的力量与魅力在于非物质性，它存在于功能、场地、业主、身体之外，在一定程度上甚至排除了尺度问题。这一抽象的普遍性是精准的"。论文发表于1996年第13届全美"基础设计学生"教学会议。

17. Art and Visual Perception: A Psychology of the Creative Eye (1954) by the eminent perceptual psychologist Rudolf Arnheim was a key influence on aesthetics of architecture in the decades of the 60's and 70's. Arnheim used science to develop a rational explanation for perception. His application of gestalt principles to works of art led to a deeper understanding of figure and ground, and its application in art and architecture.

著名感知心理学家鲁道夫·阿恩海姆所著的《艺术与视知觉》（1954）对于1960、70年代的建筑美学产生重要影响。阿恩海姆利用科学来发展对于感知现象的理性解释。将格式塔原理应用于艺术作品使人们可以深入认识图底关系及其在建筑与艺术中的应用。

18. The reader (Plan Trad | Plan Libre) was a standard photocopy booklet of about 100+ pages self-published and made available through the University Bookstore at Ohio State University beginning in 1986. It contained the course syllabus and schedule, information on the use of drafting equipment, outlines of the lectures, assignment handouts for the term, as well as readings on theory and related topics, and illustrations of works of architecture and art with diagrammatic analyses explaining formal principles.

《传统平面与自由平面》是教师编制的一本课程读物，书厚超过一百页。该读物的影印版自1986年起为俄亥俄州立大学的大学书店发行。其中包括课程大纲及日程，绘图器具的使用方式，讲座内容概要，课程作业任务书，理论及相关主题的阅读材料，建筑及艺术作品的形式图解分析。

19. The excellent introductory design manual, Creation in Space: A Course in the Fundamentals of Architecture by Jonathan Friedman, follows a similar pattern with short essays and illustrations alternating with design exercises that also confront space/form while excluding other design parameters (see "Pedagogy: Kit-of-parts Approach" for a more detailed description of the methodology and objectives). The program at the New York Institute of Technology created by Friedman was conceived at the same time as the one described in this article and shares a similar approach towards beginning design education.

乔纳森·弗里德曼所著的《空间创造》是一本出色的设计入门手册，将短篇阅读材料与插图穿插于设计练习之间，其内容专注于空间与形式并排除其他的设计因素（对于该方法及教学目标请参见"方法：装配部件教学法"一文）。由弗里德曼在纽约理工学院创造的课程与本文介绍的教学出现于同一时期，且采用相似的基础设计教学方法。

20. In "Visual Notes and the Acquisition of Architectural Knowledge" (Paul Lasseau and Steven Hurtt, Journal of Architectural Education, Vol.39, Issue 3, 1986) the connection between drawing and learning is argued. Betty Edward's theory of visualization and description of "how we draw" (Drawing on the Right Side of the Brain, 1979) is referenced and used to make a compelling case for the role of freehand drawing in the acquisition of architectural knowledge.

保罗与史蒂文·赫特所写的"视觉笔记以及建筑知识的获取"（刊于《建筑教育》，1986年第39卷第3期）讨论了绘画与学习之间的联系。该文引用了贝蒂·爱德华的可视化理论以及她在《用右脑绘画》中对于"我们如何绘画"的描述，并将此作为一个令人信服的例子来说明徒手画在获取建筑知识中的作用。

21. Albert Einstein famously once said of Le Corbusier's Modulor proportioning system, "It is a scale of proportions which makes the bad difficult and the good easy." See Le Corbusier. Peter de Francia and Anna Bostock trans. The Modulor: A Harmonious Measure to the Human Scale Universally Applicable to Architecture and Mechanics. London: Faber & Faber, 1956: 58.

在谈及柯布西耶的模度比例系统时，爱因斯坦曾指出：这个比例不容易产生丑的事物，而使得美的事物变得容易出现。

22. For a more detailed explanation of the Parallel Wall Exercise and its pedagogical value, see "Between Two Walls Lies a Space. The Parallel Wall Problem: Origins and Pedagogic Intentions" in Proceedings of the 11th National Conference on the Beginning Design Student (1994, University of Arkansas), access: http://ncbds.la-ab.com/11_Proceedings.pdf

对于平行墙练习及其教学法价值的详细讨论请参见"两片墙之间存在着空间：平行墙问题及其源起与教学法意图"一文，发表于1994年第11届全美"基础设计学生"教学会议。

23. Ulrich Franzen, Alberto Pérez Gómez, and Kim Shkapich, eds. Education of an Architect: A Point of View, the Cooper Union School of Art & Architecture (New York: Monacelli Press, 1999), 23.

插图出处 image sources

p196-1   Alexander Caragonne. The Texas Ranger: Notes from an Architectural Underground (Cambridge, MA: The MIT Press, 1995), 290.

p196-2   Cornell University, and Department of Architecture. The Cornell Journal of Architecture, vol.1(Ithaca, N.Y.: Cornell University, 1981), 58. Online access: https://issuu.com/cornellaap/docs/cja001-opt

p197   Images provided by Bruce Lonnman.

4.1-1   Roger Sherwood, Elements and Principles of Architecture, course handout at Cornell University, provided by Bruce Lonnman.

4.1-2   Colin Rowe, Robert Slutzky, and Bernhard Hoesli. Transparency (Basel; Boston: Birkhäuser Verlag, 1997), 49, 53.

4.1-3   Caragonne, The Texas Ranger, 281.

4.1-4   John Hejduk. John Hejduk, 7 Houses, Catalogue 12 - Institute for Architecture and Urban Studies, (New York, N.Y.: Institute for Architecture and Urban Studies, 1979), 82.

4.1-5   Franzen, Gómez and Shkapich, Education of an Architect, 48.

4.1-6~4.1-8   Images provided by Bruce Lonnman.

4.1-9   Cornell University, The Cornell Journal of Architecture, vol.1, 42-43.

4.1-10   Ibid., 100.

4.1-11   left: https://richardboscharchitect.com/projects/Educational_%26_institutional/Pages/Willard_State_Hospital_rehabilitation.html
              right: Sanmartin, A., ed. Venturi, Rauch & Scott Brown, (London: Academy Ed, 1986), 40.

4.1-12   Cornell University, The Cornell Journal of Architecture, vol.1, 55, 57.

4.1-13   Thomas L. Schumacher. The Cornell Journal of Architecture, vol.3, (New York: Rizzoli for Dept. of Architecture, Cornell University, 1988), 122.

4.1-14~4.1-18 Images provided by Bruce Lonnman.

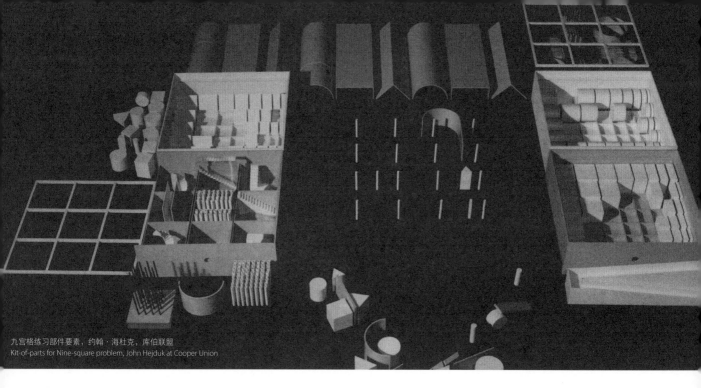

徐亮 Xu Liang

4.2 方法 pedagogy

装配部件教学法 kit-of-parts approach

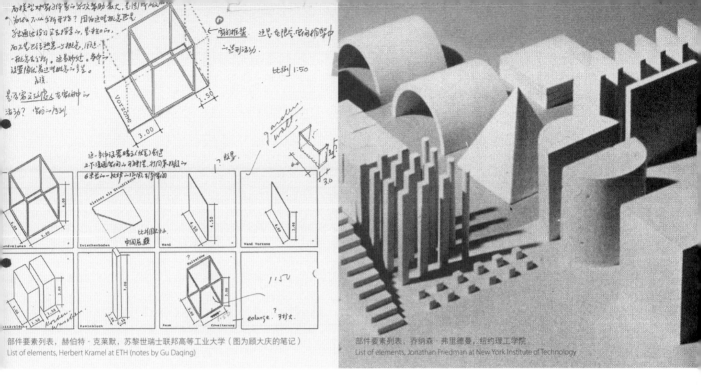

部件要素列表，赫伯特·克莱默，苏黎世瑞士联邦高等工业大学（图为顾大庆的笔记）
List of elements, Herbert Kramel at ETH (notes by Gu Daqing)

部件要素列表，乔纳森·弗里德曼，纽约理工学院
List of elements, Jonathan Friedman at New York Institute of Technology

　　建筑设计基础中的装配部件教学法，是一种针对训练空间生成与形式操作能力的设计教学方法。其突出特点在于将抽象的空间设计过程简化为给定部件要素的直观操作。以训练的评价标准作为参照，本能地摆弄一组部件要素，可以在获得可行设计方案的同时，降低初学者在面对设计问题时的盲目。同时，通过直接观察模型，可提升学生的空间感知能力。依照这一方法，无需触及功能计划、场地、建造技术等实际设计问题，可实现对于包括空间限定、虚实对比、加减法操作等在内的一系列形式问题的掌握。

　　装配部件练习的一个重要特点在于"部件要素"的设计：模数化的尺寸与几何关系可强化各要素之间的形式关系，并有效缩减设计的范围。其他的限制条件则体现在对各要素操作方式的控制。从实际的设计问题中剔除了诸多现实要素以实现对于训练的简化，这样一种抽象的方式是该教学方法中各练习的共性所在。

The kit-of-parts approach targets spatial generation and formal operations in teaching architectural fundamentals. By simplifying an abstract spatial design process into a playful manipulation of pre-designed elements, beginning design students are less intimidated by the prospect of a "blank canvas" and potential solutions may be generated intuitively with the set pieces and then evaluated against pre-established criteria. In addition, direct observation has been shown to improve students' spatial perception. With this approach, formal issues such as spatial definition, solid and void contrast, additive and subtractive process, etc., may be prioritized without additional complexity from program, site, building technology and other practical design issues.

An important feature of the kit-of-parts exercise are the constraints implicit in the design of the "parts" or elements. Modular dimensional and geometric relationships enhance the formal relationships of the elements to each other while reducing the scope of design. Other constraints govern the operational limitations of the elements. Exercises that adopt this approach accept the condition of abstraction inherent in the required simplification that excludes many realities of an actual design problem.

1. 预制装配的工业化思想

近代以来，随着工业化生产的不断发展，一些讲求高效的生产制造方式开始出现。这其中就包括预制技术：批量生产统一、标准化的建筑构件。这种装配预制构件的方法（即"装配部件"）被应用到不同的设计以及建造领域中。我们所熟悉的宜家家具[1]便是其中一例，它所倡导的"自行组装家具"就是一种部件装配思想。经"平整包装"的家具部件 [图4.2-1] 可以节省包装、储存与运输的成本，同时通过巧妙的连接设计，使得部件的安装简单易操作。

在建筑实践中，所谓"非定制的"预制建筑构件是为满足不同使用情况预先设计、生产的。对比传统的现场施工，这一做法有诸多优势：首先，大规模量产可以有效控制预制构件品质，实现产品的无差别生产；其次，在施工过程中，避免工人在现场处理繁复的施工原材料，通过直接的部件装配可加快施工进度、降低施工难度、节约施工成本。此外，这类大规模生产可以激发新的产品设计，进而推动相关产业的发展。

实际上，这种预制技术并不是一个全新的概念，它的起点可追溯到人类文明的开端。古埃及金字塔便是通过使用标准化的砌体单元（石块）来简化施工，并确保砌体墙及其基础的稳定性。在近代建筑史中，使用预制技术与标准化建造构件的例子不胜枚举。可以说，1851年建成的水晶宫是最著名"快速施工"案例。按照约瑟夫·帕克斯顿（1803-1865）的计划，水晶宫使用了模数化、工业生产的建筑框架与玻璃围护。起结构作用的铁质框架可快速架设，玻璃装配采用了温室常用的标准化单元，这一做法使得施工工期缩减至一年以内。重复结构开间的各构件组成了整个建筑系统的装配部件，并与不同层级的跨度要素、独

1. Prefabrication in industrial production

As industrial production evolved, new and more effective manufacturing methods were introduced. Prefabrication enabled the production of uniform, standardized building components. Prefabricated assemblies as a "kit of parts" have been introduced into many areas of design and construction. The IKEA "ready-to-assemble" housewares concept[1] is an example. Preconfigured furniture kits or "flat packs" [fig. 4.2-1], reduce the costs in packaging, storage and transportation. At the same time, by improving the design of the connections, end-users can easily assemble furniture pieces themselves.

In architectural practice, "off the shelf" prefabricated building components are pre-engineered and designed for use in a variety of different conditions. There are several advantages to this approach over traditional in-situ construction. First, mass production can ensure better quality and consistency in the fabrication of the components. Second, during the building process, avoiding on-site fabrication as much as possible shortens the construction period leading to cost savings. In addition, the scale of production attracts new product design and thus promotes the development of relevant industries.

Prefabrication itself is not a new concept and its origins can be traced back to the beginning of human civilisation. The Great Pyramids of Egypt

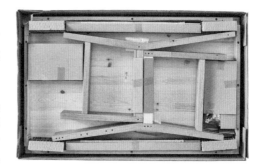

图 4.2-1
宜家座椅的平整包装；
a "flat pack" of an IKEA chair;

图 4.2-2
水晶宫典型屋架；
typical roof bay of the Crystal Palace;

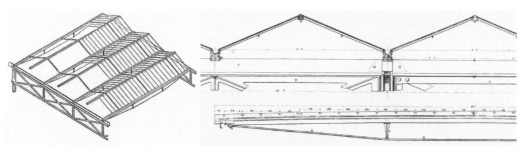

特构思的连接件一同协作 [图4.2-2]。在20世纪中期，查尔斯·埃姆斯（1907-1978）与蕾·埃姆斯（1912-1988）夫妇设计的埃姆斯住宅（1949）[图4.2-3]展示了使用标准化工业生产的建筑构件的类似方法。在此，非定制的构件（钢空腹桁架梁，厂房玻璃装配，钢楼梯等）被应用到住宅建筑中。标准的预制钢框架与模数体系（平面及立面使用的基本模数为7.5英尺，约2.28m）创造一个自由灵活的使用空间。

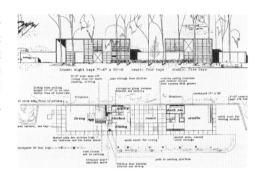

图 4.2-3
埃姆斯住宅（1949），立面及平面；
Eames House (1949),
elevation and floor plan;

2. 方法基础：幼稚园教学法

对于建筑设计教学中的装配部件教学法而言，其突出特点在于对给定部件要素的直观模型操作。这一思想在很大程度上受到了以弗里德里希·威廉·奥古斯特·福禄贝尔（1782-1852）为代表的直观性幼稚园教学法的影响。

1）"从做中学"的直观性教学思想

在讨论西方教育史的发展时，尤其是针对幼儿教育相关的理论思想，约翰·亨利赫·裴斯泰洛齐（1746-1837）与福禄贝尔是不可绕开的两位教育家。

在裴斯泰洛齐之前，学前教育的传统方式是通过授课来实现，配合课本相对理论性的介绍不同主题内容，并事先准备好问题的答案。而裴斯泰洛齐抱持的想法与此不同，他认为科学的教育方法、简化的教学过程是实现幼儿天赋能力全面发展的基础。他主张使用实物进行教学的直观方法：通过一套符合心理发展的训练，鼓励学生从直接的探索与观察中建立起对外在事物的感觉印象，并以直接经验为基础建立起内在认识[2]。在这一过程中，教师的指导也同样重要。可以说，直观教学完全改变了学前教育的传统，使教育适应自然，在教学过程中充分遵循人的内在本性以及心理发展的自然规律，同时就自然发展过程中的盲目性予以补充和改善。

在这直观教学方面，学前教育方法"幼稚园"的开创者福禄贝尔继承了裴斯泰洛齐，同时更加重视玩耍在学习过程中作用，并提出"从做中学"的观点。对于他而言，"玩耍是学习的源动力"[3]。福禄贝尔认为在幼儿的成长初期阶段，自我行为在已知与未知之间建立了联系，而已知与未知之间的矛盾促使了问题的出现，以此实现智力的增长。只有意识到了问题的存

used standardized masonry units (blocks of stone) to simplify construction and ensure the stability of the form and its foundation. In more recent history, examples of pre-fabrication and standardized building units are numerous. The Crystal Palace (1851) is perhaps the most famous example of a "fast-tracked" construction made possible by the modular, engineered building frame and glass enclosure devised by Joseph Paxton (1803-1865). The adoption of standardized greenhouse glazing units and a rapidly erected structural iron frame reduced the construction time to less than a year. The components of each repetitive structural bay constitute a building system kit-of-parts complete with a hierarchy of spanning elements and uniquely conceived connections [fig. 4.2-2]. In the twentieth century the Eames House (1949) [fig. 4.2-3], designed by Charles Eames (1907-1978) and Ray Eames (1912-1988), represents a similar approach in the use of standard industrial building components. Here, off-the-shelf components (steel open-web truss beams, factory glazing, steel stairs, etc.) have been adopted for residential architecture. The standard prefabricated steel framing and modular system (the basic dimension in plan and section is 7.5 ft) creates a flexible space for living and working.

2. Methodological foundation: Kindergarten pedagogy

In the kit-of-parts approach employed in architectural design education, the key feature is the direct operation on a set of pre-determined elements. This idea is derived from Friedrich Wilhelm August Fröbel (1782-1852) and the intuitional Kindergarten pedagogy.

1) "Learn from play" - an intuitional education idea

When discussing the historical development of Western education, especially the theory on child education, Johann Heinrich Pestalozzi (1746-1837) and Fröbel are two of the most important educators.

Prior to Pestalozzi, a conventional approach to educating children

在，幼儿才能从中获取认识。幼儿教育应当重视每个孩子的天赋与兴趣差异，指导其将这种自我行为转化成玩耍。玩耍与学习是一体的，学习可以通过玩耍来实现。此外，福禄贝尔还强调玩耍过程中规则的重要性。他认为只有在规则和限定条件被清楚认识的情况下，"不自觉的"玩耍才能够发挥最大的教学效用。

"从做中学"对于设计教学的影响是深远的，它在包豪斯教学体系也有体现。以约翰尼斯·伊藤（1888-1967）创立的基础入门课程为例，他主张学生从直接的操作（如规则限定的设计活动）中获得直接体验，并发展了"感知、理解、应用"的三步骤训练模式[4]。下文中将要讨论的装配部件教学法同样鼓励学生从设计操作的过程中发现问题，并以直观的模型操作作为解决问题的途径。

2）恩物：具有建筑学意味的幼稚园教具

在福禄贝尔之前，没有哪位教育家对于"玩耍"的教学意义有过如此细致的研究，并践行到具体教具与教学设计中。在福禄贝尔看来，伴随幼儿成长的玩具是需要经过仔细思考与设计的，并以促进其自我行为为目的。这些材料可以用来表达他们眼前的这个世界，允许他们在玩耍的过程中探索并掌握更多的未知知识。在这个思想基础上，福禄贝尔设计了一套具有教学意义的玩具——恩物[5]。

在福禄贝尔最初的教学实践以及相关文献中，恩物一套6件，依次命名为恩物一至六。入门的恩物一是一组柔软的毛线球，在色彩与质感上都与其后的恩物相异；恩物二则是一组球状、圆柱状与立方体状的木质实体，形式差异强化了针对几何形态的认识。其后的恩物是一系列以组合为目的，强调整体与部分之间关系的实体要素，其差别在于对正方体的分割方式，如恩物三将1个边长为2"（5.08cm）的立方体等分为8个边长为1"（2.54cm）的小立方体；恩物四是将这个边长为2"的立方体等分为8个长宽高比为4:2:1的长方体。恩物五的分割方式比前两种复杂，它首先将边长为3"的立方体等分为27个边长为1"的小立方体。然后将其中6个小立方体沿对角线方向分割，得到6个1/2的三棱柱以及12个1/4的三棱柱。在恩物六中，边长为3"（7.62cm）的立方体被分割成3种比例不同的长方体。［图4.2-4］

从恩物中，我们可以看到一些与建筑设计教学关

consisted of a lecture on the subject at hand, including theory, with reference to books, and the offering of ready-made answers. However, Pestalozzi believed that a scientific educational method and a simplified teaching process could better contribute to the actualisation of children's latent talents and faculties. Most important, by constructing a series of psychologically ordered exercises, direct exploration and observation through their activities are primary in establishing an impression of outward objects, on which an intuitive understanding can be generated with adequate stimulation and enrichment.[2] This intuitional pedagogical idea dramatically transformed the education tradition. In this way, education can harness children's natural innate learning ability with minimum interference, and at the same time, avoid the aimlessness of unstructured learning.

Fröbel, the creator of the pre-school educational method called "Kindergarten" (garden of children), carried forward Pestalozzi's educational thinking and proposed the concept of "learn from play". For him, "play is the engine of real learning"[3]. In early childhood, self-directed activity allows a child to establish a connection with the physical world and confront contradictions between the known and the unknown. These self-activities can foster intellectual growth. Children can obtain knowledge once they have recognized a problem. Early-childhood education should pay attention to the differences in individuals' talents and interests, and accordingly provide instruction to transform the self-activities into play. In this way, play and learning may be integrated. Fröbel specifically emphasizes the importance of rules within play: only when the rules and limits are precisely understood, unconscious play can achieve its maximal function.

"Learn from play" has had a profound impact on design education, as evidenced in the Bauhaus education system. For example, the Basic Course, established by Johannes Itten (1888-1967), required obtaining direct experience from an operation (such as a design activity guided by rules), based on which he developed a three-step training model - perception, comprehension, and application[4]. The kit-of-parts approach, to be discussed below, also encourages students to identify problems within a design process, and then solve them in a similar manner.

2) Gifts: Kindergarten educational tools with architectural implications

Before Fröbel, the significance of play in education had not been explored in detail and or applied to the design of teaching tools. It was Fröbel's belief that toys accompanying childhood should be thoughtfully considered and designed in order to stimulate self-activities. These toys would lead children to interact with the world in a guided but

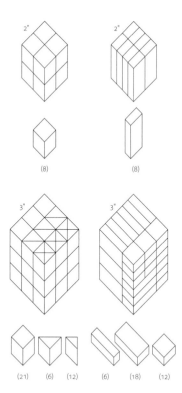

图 4.2-4

恩物三至六，
体块分割与单元列表；
Gift 3-6,
block division and list of
elements；

2" 2"

(8) (8)

3" 3"

(21) (6) (12) (6) (18) (12)

unobtrusive manner, encouraging discovery and acquisition of new knowledge. Based on this thinking, Fröbel created a set of educational toys he called "gifts"[5].

The gifts in Fröbel's text, *Pedagogics of the Kindergarten* (1904), are named Gifts 1-6. Gift 1 consists of six yarn balls, which are different in colour and texture from the following gifts; Gift 2 is a set of wooden solids, which enhances the understanding of different shapes, i.e. sphere, cube, and cylinder. The design of the other four gifts is based on the concept of combination, with an emphasis on cube subdivision and the relationship of the parts to the whole. Gift 3 divides the 2" (5.08cm) cube into eight 1" (2.54cm) cubes, while Gift 4 divides the same cube into eight rectangular prisms with a proportion of 4:2:1. The division of Gift 5 is much more complex, in which a 3" (7.62cm) cube is first divided into twenty-seven 1" cubes. Next, five of the twenty-seven cubes are divided diagonally into triangular prisms, six half-cubes, and twelve quarter-cubes. In Gift 6, the 3" cube is divided into rectangular prisms with three different proportions. [fig. 4.2-4]

Fröbel's gifts embody concepts that can easily translate to architectural design education. At the outset, one gains an understanding of the relationship between the "part" and the "whole" by disassembling the whole cubes (Gifts 2-6) first, and when the play is done, reassembling the pieces before returning back to the box. All elements must be incorporated. Secondly, the difference in the form and quantity of the elements is made evident from the various subdivisions of the cubes. In Gift 4 a discussion on proportion (e.g. 1:2:4), module and orientation, is introduced by the rectangular prisms. Gift 5 introduces the non-orthogonal idea "angle", by the diagonal division. With the added elements in Gift 4 and 5, it is possible to configure more complex forms, still abstract but tending more towards *representational*. These collective assemblies are compositions of forms organized by formal ordering principles, e.g. symmetry, balance, hierarchy, etc. In addition, the concept of process is embedded: a creation can be modified and transformed through the reordering of the elements. In the process of this kind of interactive play, communication is key for connecting perception with the language system. On this aspect of the relationship of the gifts to formal organisation in architecture, Frank Lloyd Wright (1867-1959) stated: "… along with the gifts was the system, as a basis for design and the elementary geometry behind all natural birth of form… I soon became susceptible to constructive pattern evolving in everything I saw. I learned to 'see' this way and when I did, I did not care to draw casual incidentals of Nature, I wanted to design."[6]

联的想法。首先是整体与个体之间的关系：在操作之前，幼儿需要从盒中整体地取出恩物（二至六），并在使用后需恢复到其原始形式并整体放回盒中。在操作过程中，应使用该恩物的全部要素。其次，由于各恩物的分割方式不同，导致出现要素间形式与数量的差别。在恩物四中，长方体要素引入了对于比例（如1:2:4）、模数以及摆放方向的讨论。在恩物五中，因为对角线的分割，还带来了关于"角度"的问题。而恩物四、五中，要素数量的增加为创造复杂的形式组合带来了可能，虽然形式是抽象的，但具有表现能力。对于要素的组合，即是形式以及形式秩序原则（包括对称、平衡、中心等）的构成。另外，重视对于训练过程的控制，搭建的过程应当处于不断的调整与发展中，不可直接推倒重来。在这个重视互动的训练过程中，沟通是在感知与语言词汇系统间建立联系的关键。而恩物与建筑形式建构之间的互通，弗兰克·赖特（1867-1959）曾说："恩物是一个系统，它是设计以及所有形式生成的要素几何的基础。很快的，开始对事物内在的结构性图案变得敏感。我学着去观察它的生成方式，这个过程中并不是对自然现象的描述那么简单，而是开始了设计。"[6]

3. 建筑设计教学中装配部件教学方法的实践

在此讨论的"装配部件教学方法"是建筑基础教育中针对于空间感知与操作能力培养的一种教学思路：在明确的限定条件下，通过操作一组给定的模型要素部件，完成针对建筑空间问题的设计训练。它是20世纪70年代到21世纪初，美国建筑院校基础教学中最为常用的教学方法之一[7]。以约翰·海杜克及其"九宫格问题"为代表的"得州骑警"教学实验被认为是装配部件教学法在现代建筑教育中实践的起点[8]。

事实上，在正规职业建筑师培养教育体系的初端——法国学院派建筑教育中，装配"空间"的思想已经存在。以19世纪早期法国最重要的建筑理论家与教育家让·尼古拉斯·杜朗（1760-1834）为例，在《综合工科学院建筑学课程概要》（1823）一书中，他将建筑的建构分为三个层次：建筑要素、建筑部件与构图组织。在这个体系中，基本的建筑要素包括墙体、壁柱、拱券等，通过组合建筑要素可形成建筑部件，包括门廊、前厅、房间等[9]。值得注意的是，中间层次的建筑部件本质上就是一个个具有功能的空间单元。建筑设计便是将这些空间单元通过轴线进行串联。柱间距作为构图的基本模数单位，在这个设计过程中控制各建筑部件间的连接。在进行建筑装配的过程中，不同部件间的大小关系可以用数值进行表示。

1）一套全新的建筑教学课程：得州骑警

以科林·罗和伯纳德·霍伊斯里为代表的"得州骑警"（一群于1954-1958年间于得州大学任教的年轻教师）对于现代建筑教育的突出贡献主要体现在教学方法上[10]。1954年，他们提出了一套修改过的课程教学计划，该计划的核心来自于针对现代建筑中的"空间"所下的更准确定义。从对于一批建筑大师的重要作品分析开始，科林·罗与霍伊斯里辨析了现代空间的主要特点，以及区别现代建筑的关键特征。在此之前所发表的一篇重要文献——"理想别墅的数学"（1947）中，科林·罗对佛斯卡里别墅（帕拉第奥，

3. The kit-of-parts approach in architectural design education

The kit-of-parts approach discussed here is a pedagogical method in the teaching of spatial perception and operation in architectural beginning design. By manipulating a given set of model elements, students are freer to focus on specific spatial issues in a design problem. This approach has been dominant in the basic design teaching at American universities from the 1970's on[7]. The Texas Rangers' teaching experiment, exemplified by John Hejduk's nine-square problem, may be seen as the first instance of the application of a kit-of-parts approach in contemporary architectural education[8].

It should be remembered that the idea of "assembling space" has existed since the origin of formal architectural training in the French École de Beaux-Art. In the *Précis des leçons d' architecture: Données a l' Ecole impériale* (1823), Jean Nicolas Louis Durand (1760-1834), an important French architectural theorist and educator in the 1800s, divided building tectonics into three levels: elements of a building, parts of a building, and composition. In this system, the fundamental elements include walls, piers, arches, etc. By combining these basic elements, parts of buildings (e.g. porches, lobbies and lounges) are produced.[9] It is important to note that the parts are modular spatial units with specific functions. Architectural design here is the composition of units, or parts, along axes. In this case, the basic module is the *inter-axis*, the distance between the centerlines of two columns. It is essential in the assemblage of a building since the relations of size among the parts can be expressed in whole numbers.

1) A new curriculum for architectural education: the Texas Rangers

A major influence on pedagogy in architectural education in the mid-twentieth century was the contribution of the Texas Rangers, a group of teachers at the University of Texas (1954-58) led by Colin Rowe and Bernhard Hoesli[10]. In 1954 they proposed a revised teaching program that at the core, was guided by a more precise definition of space in modern architecture. Beginning with research on the important works of a few leading modern architects, Rowe and Hoesli identified the main characteristics of modern space, and the principal features that distinguished modern architecture. In an important essay previously published ("The Mathematics of the Ideal Villa", 1947), Rowe compared two buildings, the Villa Foscari (Andrea Palladio ca.1550) and the Villa Stein (Le Corbusier 1927-28), pointing out the common proportions and subdivisions of the plan organization of the villas, while at the same time noting the differences in spatial character, structure, and composition. This established a precedent for the formal analysis of works

约1550）与斯坦因别墅（勒·柯布西耶，1927–1928）进行比较，同时指出两者在空间特点、结构与构图上的差异，以及在平面组织的比例及分割上的趋同。这为后来对不同历史时期与风格的建筑作品进行形式分析提供了先例示范。其后，在《透明性》（1963）一文中，科林·罗与罗伯特·斯洛茨纪应用格式塔理论来分析现代绘画与建筑，提出一种不同的"透明性"概念以暗示二维与三维情况下空间图形的相互渗透与重叠。这些文章反映了当时对于现代建筑的不断研究与分析，对于同期得州大学的设计教学有很大的影响。这一研究在基础训练的诸多设计练习中都有体现。

"九宫格问题"的雏形是斯洛茨纪与霍伊斯里在视觉绘画课程中设计的一个三维空间训练：在一个由立方体组成的3格 x 3格的网格中，沿网格线布置一定数量的灰纸板，以围合、限定、分割整个平面空间布局。训练的重点在于讨论板片之间塑性的张力与压力[11]。缺乏建筑性一直都是这个训练的缺陷，直到海杜克将其与框架结构关联，才解决了这一问题。将立方体之间的间隙发展为垂直向的柱，并通过梁来建立柱之间的水平联系，这样便出现了一个梁柱框架结构[图4.2-5]。我们可以从以下角度来解读这个练习：底板可以被认为是基地平面，通过梁可支承水平的屋面板片，垂直纸板则置于其中用以限定空间。从形式的角度，框架结构为构成提供了基本参照，通过摆放墙体等部件要素，可以重新组织中性的框架空间，并创造具有层级差异的复杂空间关系。

在"九宫格问题"中，初学者开始关注于抽象的空间设计问题。看到这一训练的潜力所在，一年级的基础设计与二年级的建筑设计入门发展出了一系列的类似练习，从练习要求中可以看到它们在作业设计上的共性：给定场地及其尺寸，明确数量、形式及其尺寸的操作要素，具体的设计任务[12]。装配部件方法也成为了这些"每日一题"练习中所贯彻的思想，在专注于单一特定设计问题的同时，确保训练的复杂程度可控，允许学生有时间来开展其他活动。

在离开得州大学之后，1954至1957年间的这批年轻教师将这一教学法思想带去了其他学校。在苏黎世瑞士联邦高等工业大学的一年级设计课程中，霍斯里继续发展了一套基础设计训练[13]。这套教学由一组小练习开始，并使用带有装配部件思想的一系列空间限定

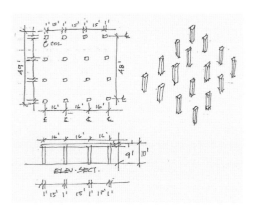

4.2-5

九宫格问题：框架结构，约翰·海杜克；
Nine-square problem: frame structure, John Hejduk;

of architecture of different historical periods and styles. Later, applying the principles of German gestalt theory in an analysis of modern painting and architecture in "Transparency: Literal and Phenomenal" (1963), Colin Rowe and Robert Slutzky suggested an alternative meaning of transparency that implied the interpenetration and overlap of spatial figures in both 2D and 3D. These published articles reflect the on-going research and study of modern architecture that influenced the design teaching at the University of Texas during this period. Many of the design exercises in the foundation year programme were instigated by this research.

An early version of the "nine-square problem" was a three-dimensional exercise developed by Slutzky and Hoesli for an introductory drawing course. "A grid work of nine equal cubes, three units wide by three units deep, into which a given number of panels of grey cardboard could be inserted on edge. These panels could then be arranged so as to enclose, define, and divide any number of elementary spatial configuration. (The exercise aims to discuss) the plastic extension and compression of planes"[11]. The lack of an architectural reference had been somewhat of an issue for this exercise until Hejduk drew parallels with a modern two-directional frame structure. When the interstices between the nine cubes are developed into vertical columns to be connected by horizontal beams, a post and beam structure is generated [fig. 4.2-5]. We can interpret this exercise in the following way: model base as site; beams to support the horizontal roof panels; vertical planes used as partitions to define the space within the frame, etc.[12] In terms of form, the grid frame structure is a datum in the composition. By introducing pre-configured elements (flat planes) into the structural framework, it is possible to reorganise the neutral framed space, and achieve a complex spatial relationship with hierarchy.

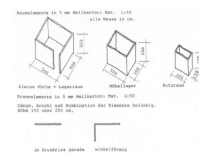

图 4.2-6
给定空间限定要素，
伯纳德·霍斯里；
given space-defining
elements, Bernard Hoesli;

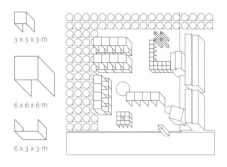

图 4.2-7
基本空间体块与场地组
织，赫伯特·克莱默；
basic architectural units
and site arrangment,
Herbert Kramel;

要素［图4.2-6］。在其继任者——赫伯特·克莱默的建筑入门设计课程中，使用了由卡纸折成的门字形"基本空间体块模型"（即基本空间单元）来构思场地布置［图4.2-7］；就具体的空间设计训练，利用网格来组织一系列模数化的墙体、柱等空间限定要素，部件要素的形式多种多样[14]。与此同时，沃纳·塞利格曼、李·祝辰等人在康奈尔大学的基础设计课程中也发展了类似的练习，要求在限定的范围里组织给定的部件要素。

2）乔纳森·弗里德曼《空间创造》

自20世纪80年代中期起，乔纳森·弗里德曼在纽约理工学院的建筑学院开设了一门基础设计课程。弗里德曼毕业于库伯联盟，他曾经是海杜克的学生。在纽约理工，他需要在35名讲师的协助下，每学期教授3个校区共700多名学生。这就需要一套方法，使得教学过程持续且不受指导老师的个人差异影响。在《空间创造：建筑学基础课程》（1989）一书中，弗里德曼详细阐述了该课程的架构方式、具体的训练内容、学生习作、相关的阅读材料与参考资料。这套书分上下两册，第一册名为"建构"，其中包括7组相互关联的练习，依次名为：统一、对话、容积、变形、表现、时计及自由创作[15]。每组练习又分为三维的空间建构训练与二维图形训练两部分。这些练习由简易及复杂、由抽象及具象，循序渐进依次展开。

表2罗列了课程的前三组练习以及它们所使用的材料，即部件要素。练习一以随手可得的树枝与石头作为材料，要求学生以此来创造一个独特而统一的形式。在其后的各练习中，操作的部件要素由自然形向具有模数关系、规则的抽象几何形转化。在练习二

Inspired by the potential of the nine-square problem to allow beginning students to focus on issues of space design in the abstract, Introduction to Architectural Design (Year 1) and the Sophomore Design Studio also developed a set of similar exercises[13]. From the exercise briefs, we can observe their common features: a field with specific dimensions, a set of given elements (quantity, form, and dimension), and a detailed design task. The kit-of-parts approach is a theme throughout these "lesson of the day" assignments, facilitating a focus on a particular design issue while keeping the complexity of the work manageable, thus leaving time for other activities.

After leaving UT during the years 1954-1957, the teachers involved transported this pedagogical thinking to a number of other schools. At the ETH in Zurich, Hoesli pursued a version of the curriculum in the first year Grundkurs (basic design) course. Beginning a sequence of exercises, students first generated an element of spatial definition [fig. 4.2-6], an embedded kit-of-parts concept. And in the teaching of his successor, Herbert Kramel, "basic architectural units" made by cardboard "loops" (hollow box elements) were introduced to organise the site [fig. 4.2-7], while in the exercise of defining architectural space, a set of modular elements, e.g. walls, columns, were assembled on a grid base[14]. Werner Seligmann, Lee Hodgden, et. al developed similar exercises of limited scope and incorporating given elements, concurrently in the basic design program at Cornell University.

2) *Creation in Space* by Jonathan Friedman

Beginning in the mid-1980's, a beginning design program at the School of Architecture at the New York Institute of Technology was developed by Jonathan Friedman, himself a graduate of the Cooper Union and a former student of John Hejduk. Teaching about 700 students from three campuses with the assistance of 35 lecturers every semester, Friedman

Ex.1 统一 | UNITY	Ex. 2 对话 | DIALOG		Ex. 3 容积 | VOLUME

表 2

练习1-3及其部件要素，
乔纳森·弗里德曼；
exercise 1-3 and the
kit-of-parts (elements),
Jonathan Friedman;

a.

b.

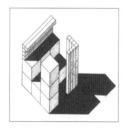

12根木棍：
长15.24cm，直径3~6mm
12颗石头：每颗重56~85g

12 sticks: 6" x 1/8"-1/4"ø
(15.24cm x 3-6mmø)
12 stones:
2-3oz. (56-85g) each

12根木棍：
长15.24cm，直径3~6mm
12颗石头：每颗重56~85g
1块底板：20.32cm x 20.32cm

12 sticks: 6" x 1/8"-1/4"ø
(15.24cm x 3-6mmø)
12 stones:
2-3oz. (56-85g) each
1 base: 8" (20.32cm) square

12个立方体：边长2.54cm
12根细棱柱：截面0.635cm x
0.635cm，长7.62cm
1块底板：20.32cm x 20.32cm

12 cubes: 1" x 1" x 1"
(2.54cm x 2.54cm x 2.54cm)
12 rods: 1/4" x 1/4" x 3"
(0.635cm x 0.635cm x 7.62cm)
1 base: 8" (20.32cm) square

12个立方体：边长2.54cm
12根细棱柱：截面0.635cm x
0.635cm，长7.62cm

12 cubes: 1" x 1" x 1"
(2.54cm x 2.54cm x 2.54cm)
12 rods: 1/4" x 1/4" x 3"
(0.635cm x 0.635cm x 7.62cm)

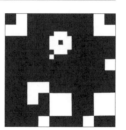

中，学生需用使用一组给定的要素（包括一块模型底板）来探索中心与边缘、平行与对角、水平与垂直、体量与空间、实体与虚空的"对话"。这些辩证的概念会根据形式操作法则发生变化。在这个过程中，还有引入其他问题的讨论，如场地，部件要素本身的性质差异（包括大小、形状、比例等）。练习三可以看作是练习二的延伸，在这个全新的练习中，学生需要使用与练习二相同的部件要素来讨论不同的问题：通过堆叠给定的实体要素来限定—个边长为3''的立方体体量。由于部件要素的体积之和小于规定的立方体体量，这样实体之间势必会余留一些空间，这样便产生了针对实体与虚空之间关系的讨论。在完成设计后，学生通过绘制平面、剖面图解以继续讨论二维层面上的图底关系 [图4.2-8]。

通过这些短周期的设计练习，该课程突出强调以

needed a methodology that would provide a consistent teaching approach independent of the individual instructor. In *Creation in Space: A Course in the Fundamentals of Architecture* (1989), Friedman elaborated such a course structure, with design exercises, examples of student works, together with relevant reading and reference materials. The first book of the two-volume set, Architectonics, details seven pairs of connected exercises. These are Unity, Dialog, Volume, Transformation, Expression, Timepiece, and Free Exercise[15]. For each pair, there is a two-dimensional graphics exercise and a three-dimensional "architectonics" model design project. These exercises are arranged sequentially, and grow increasingly complex, moving from the abstract to problems of a more realistic nature.

Table 2 displays the first three exercises together with examples of the assignments and a list of required elements (the kit-of-parts for each assignment). In Exercise 1 students gather sticks and stones from nature and make a 3D composition exploring how the two types of materials, stones and

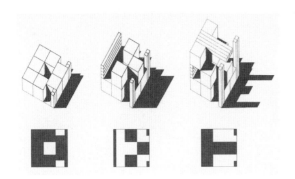

图 4.2-8
实体与虚空，
图底模棱两可；
solid and void,
F/G ambiguity；

下4方面内容：（1）场地：重视认识场地本身具有的形式特征（如常见的方形），以及它对于其上部件要素组织的影响。这些性质或形式关系可以激发对应于组织策略的视觉状态。这一点还可以通过组合要素构成及其与场地的位置关系得到加强。（2）比例：对于初学者而言，比例是一个相对难懂的概念，但是它对于组织一个统一的构成至关重要。尺寸模数系统是比例的突出特征，该课程所使用的部件要素均以1/4''（约0.635cm）为基本模数单元。（3）对比：通过水平与垂直、开口与闭合、正交与对角、内与外，图与底、实体与虚空的对比，可以创造强烈的识别性与差异性。若要获得一个复杂而有序的构成，"统一"与"多样"均不可或缺。（4）变形：形式变形的方式可分为两类，一是改变部件的位置与数量，如复制、序列、分层等；二是在改变部件组织的模式，如增加、挖去、相连、相交等。通过一系列有序的动作，可生成一个复杂的形式组织。

在弗里德曼的教学中，空间的感知训练并不依赖于传统教学中相对抽象的学习模式，而是鼓励"从做中学"、强调对于模型部件要素的直观操作：在"统一"与"对话"练习中感觉不可见空间的存在；理解空间作为虚空的体量，它的形状应当准确、不含糊。当它的各边、角及界面在三维中相互对位时，空间的塑造也会相应简单。事实上，视空间为一种有形状有尺寸的形式是建筑学的一项基本能力，它有助于提升学生对于空间及其变形的敏感程度。加上对图底关系、现象的透明性等概念的讨论，学生能够就空间形式进行阅读与分析，并为空间设计提供新的想法。

sticks (tree branches) create a unique form. Thereafter, the elements used in the exercises shift from organic shapes to modular and abstract regular geometric forms. In exercise 2, using the given elements (including a model base), students investigate the dialog between centre and edge, orthogonal and diagonal, horizontal and vertical, mass and volume, and solid and void. These dialectic pairs are transformed through principles of formal operation. During the process, other issues are introduced, such as site, as well as the differences in the intrinsic properties of the elements, e.g. size, shape, proportion, etc. Exercise 3 proposes a new design problem that is an extension of Exercise 2: define a 3″ cubic volume by stacking the solid blocks created in exercise 2. However, the total volume of the elements combined is less than the volume of the 3″ cube, thereby making the presence of voids in the larger cube unavoidable. These leftover voids seen in relation to the solid portions of the cube initiate a discussion of solid-void relationships. This is then further examined in 2D by the figure-ground relationships revealed in plan and section cuts [fig. 4.2-8].

In these short design exercises, the course addressed four key issues: (1) The *field* or site. Identifying the formal characteristics of the field (a square in most instances) and its geometric properties. These properties or formal relationships generate visual conditions that suggest organizational strategies. This can be reinforced by the configuration of the composition and its position relative to the base. (2) *Proportion*. For beginning students proportion is not well understood yet it is a key aspect of a unified formal composition. The dimensional module is a critical feature of proportion. The kit-of-parts used in this course is based on a 1/4″ (0.635cm) basic modular unit. (3) *Contrast*. The oppositions of horizontal and vertical, open and closed, orthogonal and diagonal, inside and outside, figure and ground, solid and void, are essential in generating differentiation and identity. A complex yet ordered composition requires both consistency and variation. (4)

4. 作为一种教学手段的装配部件方法

通过上文的分析，我们可以清楚地把握装配部件教学方法的思想源泉、方法基础以及它作为建筑教育中的一种教学策略的实际应用。下面我们可以从4个方面对该教学方法的实践进行总结说明：

1）部件要素：形式、数量与尺寸

在任何装配部件教学法的实践中，确定各部件要素的形式、尺寸以及数量都是必要的。对于基础设计教学而言，对要素进行抽象、与现实中的物件保持一定距离同样关键。这些要素可以按照形状分为三类：体块（立方体或长方体形式，其三个向度上的尺寸比较均衡）、杆件（或长棱柱，其某一向度的长度比另外两向度突出）与板片（某两向度的尺寸较第三向度突出）。实际上，这些形式暗示类比建筑的作用：立方体的要素是积聚的实体体积，倾向于独立支承存在；杆件则突出于它的线性特征，接近于建筑中的梁、柱；而板片常用于表示墙体或楼板构件，可利用其表面进行空间限定。

设计部件要素时，对于其尺寸的考虑尤为关键，需要就不同要素的模数及比例关系进行协调。要素的整体尺寸及比例尺通常取决于两方面因素，其一是预设的模型比例尺，这要求模型要素在经过比例换算后，其尺寸满足一般建筑空间在使用上的基本要求。尽管模型本身是抽象的，甚至有时还会刻意与真实空间、功能保持距离，但是诸如墙面洞口等在内的动作还是会暗示一定的人体尺寸。另一方面考虑则是可获得的模型材料尺寸。在制作装配部件各要素时，应避免造成浪费以及不必要的人工，这就需要就原材料的尺寸来调整各要素的大小以及相应的比例尺。比如说模型常用的木条直径为1"，如果我们将训练中使用的圆柱体要素的直径定为24mm，其制作便会相对简单。

部件要素的数量、奇偶差异往往会为设计练习带来一定的挑战，继而影响到关乎构成的一系列决定。如果部件要素完全采用偶数数量，往往容易出现形式对称的设计结果，这便限制了其他形式可能的出现。同时，某一特定要素的数量会直接影响它在构成设计中的作用。举个例子，如果我们希望设计中有一个非常重要的要素，它是整个设计的焦点，那就它的数量应当设定为"1"，如1个立方体或1个圆柱体。

Transformation. There are two approaches to formal transformation proposed. One way is to change the position or number of elements through grouping, sequencing, layering, etc. The other way is to alter the pattern through addition, subtraction, union, intersection, etc. The implication is that a complex formal organisation may be achieved through a guided series of actions.

For Friedman, the development of a spatial perception relies on "learning from play", not conventional and more abstract forms of learning. Through the direct assembling of the model elements, students are able to feel the presence of the "invisible" space in the exercises Unity and Dialog; understanding the forming of space as a void volume whose shape is precise and unambiguous is much easier when seeing the alignment of edges, corners, and planes in 3D. The ability to visualize space as a form with shape and dimension is a fundamental skill that promotes the development of a heightened sensitivity to space and its transformation. With the addition of figure/ground representation and the concept of phenomenal transparency, one gains additional tools that contribute to the reading and analysis of spatial form, offering new insight into spatial design.

4. The Kit-of-parts approach

From the above discussion, we gain understanding of the intellectual origins and methodological foundation of the kit-of-parts approach, as well as some of the instances of its use as a teaching strategy in architectural education. From the perspective of teaching practice, we can summarise the pedagogy in the following four aspects:

1) The Elements: form, number and dimension

In any kit-of-parts, it is necessary to determine the forms, dimensions and quantity of the elements, the so-called parts of the kit. In beginning design education, these elements are intentionally abstract forms lacking in association to real objects. In terms of shape, the elements tend to be of three types: block (cube or cuboid forms in which the three dimensions of length are similar), stick (or rod, in which the dimension in one direction is larger than the other two); and slab (the dimensions at two sides of the object are much longer than the other). The forms suggest analogous architectural functions: cuboid elements are compact solids and tend to act as free-standing objects; sticks recall column or beam elements in their linearity; and slabs resemble wall or floor-like elements, acting as space definers because of their planar surfaces.

Dimensional consideration is important in the design of the elements, and requires a certain coordination of dimensions to enhance modular and

最后，统一各要素的模数关系会为构成设计带来一些预料之外的便利，如各要素在尺寸上的相互匹配。其中的一种做法是使用选定尺寸的整数倍尺寸，如6mm、12mm、24mm等。这样在构成中，它们之间有较好的对位关系。当多个要素组合在一起时，还可获得尺寸更大且具有同样模数关系的要素。对于模数的考虑，还可以延伸到模型底板或场地的尺寸，这样既可以对场地进行模数划分，也可以更配合其上各部件要素的布置。

2）规则与条件

明确规定部件要素的使用与操作，是装配部件教学法的一项重要原则。通常情况下，就数量而言，应在构成设计中使用全部的给定要素。当然也存在一些可相对灵活处理的部件要素，它们可以为设计带来更多的创意可能。这一类的典型要素有分隔墙体板片。用于空间限定时，可进一步变化它们的形式，如L形墙、U形墙等。通常这些墙体的厚度与高度是规定的，而它们的长度可变。在训练中，往往不会限制这类要素的数量。这一点与它们本身的特征有关：当板片以很小的间距积聚在一起时，会创造出构成中相对密实的区域。

在任何设计训练中，都潜在诸如构造、材料、功能计划、尺度关系的因素。它们在模型中可表现为具体的操作方式限制，比如说墙面上洞口的高度与宽度。洞口的宽度往往取决于其上过梁的跨度，其长度与高度关系需要满足结构需求。对于门洞而言，可通过限定其高度来暗示基本的人体尺度关系。在建筑中，"光照"是另一个不可避免的因素。为防止屋顶覆盖面积过大对室内自然采光造成的影响，设计练习应限制水平屋面板片面积百分比。恰是这些部件要素的具体特征以及它们在使用中的限制条件，使得装配部件训练区别于其他的抽象空间构成练习。

以四弦乐器为例，虽然弦的数量有限，它同样可以演奏出无数的美妙乐章。我们相信，在装配部件教学法中，明确具体的操作方式并不会减少设计的可能性。相反，它能有效避免出现盲目的设计操作——如果设计中什么都是可能的，反倒会丧失清晰的发展方向。当学生面对于一个有着明确目标以及清楚界限的设计问题时，其创造力可以得到最大程度的激发。空

proportional relationships. The overall size or scale of the elements is often based on two parameters. The first is an assumed scale of the model. The elements in general have a dimensional range that fits the assumed scale of the model as an architectural (habitable) space. Although the model is abstract and may not be intended to represent a real place or function, openings in walls with lintels, for example, begin to suggest a human scale. The second parameter is the available dimensions of the materials used to create the elements. To minimize waste and unnecessary labour in forming the pieces of the kit-of-parts, it sensible to adjust the size and scale of the elements to fit the dimensions of materials from which it will be made. A wooden dowel typically is available as a 1" diameter rod, therefore a round cylindrical element is more easily created if its diameter is 24mm.

The quantity of the elements, and their odd-even differentiation, increases the challenge of an exercise and subtly affects decisions about composition. If the elements are all in even number quantities, the tendency may be to produce a symmetrical organization scheme and thus limit other formal possibilities. The quantity of a particular element may be directly related to its role in a design. For example, if a hierarchical and more dominant focal element is desired as a visual focus, a singular unique element is preferred (a cube or a cylinder).

Finally modularity of the elements will have unintended benefits on a composition through the natural matching of dimensions. Multiples of whole numbers in the selection of dimensions (6mm, 12mm, 24mm, etc.) will make it easier to create alignments and to combine elements together to form larger modular assemblies. The modularity might also extend to the dimensions of the base or field. This will produce modular subdivision of the field and instigate a more natural fit between the elements of the kit-of-parts and the site.

2) Rules of Engagement

Providing restrictions on the deployment and operation of the elements is an important principle in the kit-of-parts approach. In general, all of the formed elements of the kit-of-parts should be used in a composition. There can be flexible or "wild card" elements to give the designer some additional creative control. One such element is the non-specified partition wall. These secondary space-defining elements may take on different configurations (L-shape, U-shape, etc.). Their thickness and height is specified but the length can vary. The quantity of these elements may also remain unspecified. Because of the unrestricted nature of these elements, they have great potential as infill, creating more densified areas in a compo-

白的画纸往往会成为设计的最大难题，它并不会指向创造性的方向。

3）循序渐进的学习方式

在装配部件教学法中，对于形式构成设计的掌握需经历一个循序渐进的训练过程。教师通过设计各练习及其需要使用的部件要素，可以在明确设计任务的同时强化学生的学习。这样一种循序渐进的教学方式要求我们将关于形式构成的知识系统分解为一个个的单一问题，然后通过设计一套练习使得其中每一步探索都针对于某一新的想法或概念，并在前一练习的基础上持续积累。在每一步中引入新的问题，使得练习逐渐复杂化，学生也相应利用更多的时间来生成及发展一个合宜的设计方案。

在这个学习过程中，装配部件就好像是一副脚手架，在不断深入的设计问题之间创造连续性。设计的初始不再是一张空白的画纸，在他们的手边有一套形式多样的物件，通过试错的方式对其进行摆弄、组织。每一次的尝试都会带来新的可能，同时也提出了新的问题。这一点，可以说是装配部件教学法最为突出的优势。通过引入限制条件来促进设计决策，以此了解什么是不可行的，并在一个相对明确的可行范围内进行探索。

在抽象形式构成的知识系统中，包括了诸多由霍斯里、科林·罗所建立的空间感知、空间设计概念，这当中比较突出的概念是"将空间感知为具有形状与尺寸的形式"。"空间渗透"以及"现象的空间透明性"概念为理解现代建筑构成奠定了理论基础。对于图与底的区分进一步强调了我们对于物体与场地之间关系的理解，并关联空间构成的形式语言。在教学中，可以通过这一系列渐进的、问题明确的设计训练来介绍这些概念，并允许学生在练习中探索。

4）教学法特点

值得注意的是，装配部件教学法在设计过程中有几点突出特征。第一是对于直观操作的重视。装配部件教学法鼓励直接操作预先设计过的模型部件要素。通过摆弄这些物件要素，学生可以获得对于形式的触觉与视觉认识。通过与形式的直接互动，学生可以对正在发生的设计产生直观的三维印象。对于空间设计

sition by increasing their number.

In any design exercise, issues of construction, materiality, program, scale, etc. are subliminally present. They are implied by the form of the model reflected in the detailed limitations of every operation, for example, the height and width of a wall opening. Through passage is suggested by the lintel spanning the opening, the length and depth of which conforms to structural norms. Likewise the height of the portal opening implies a human scale figure. Another condition that is unavoidable in architecture is light. To prevent roof covered spaces from becoming too large and thereby preventing natural light from entering, the percentage of horizontal covering may be specified. These specific characteristics and restrictions on elements in the kit-of-parts exercise distinguish it from other abstract exercises in spatial composition.

A musical instrument like a violin has only four strings yet it can create an infinite variety of music. Similarly, we believe that the specifications and restrictions that are part of the kit-of-parts approach do not limit design possibilities. Instead, the opposite is more likely; if everything is possible, there is no clear direction forward. Creativity works best when faced with a problem that has a clear objective and boundaries. The blank canvas is often more of a barrier than an invitation to creativity.

3) Progressive learning

Acquiring knowledge of formal composition in design by a step-by-step process is greatly enhanced with a kit-of-parts approach. Each design exercise can be focused on one or a few issues and the kit-of-parts associated with the exercise can be designed to strengthen learning within the defined boundaries of the problem. Progressive learning breaks the knowledge content of formal composition into singular issues and creates a set of exercises each specifically crafted to explore a new idea or concept and build on the learning of the previous. As more issues are introduced in each step, the exercises become progressively more complex, requiring a longer period of time to generate and develop an appropriate design scheme.

In this learning process the Kit-of-parts acts as a type of scaffolding, providing some continuity between successive problems5. Rather than beginning with a "blank canvas", the student has at their disposal a set of forms to arrange by trial-and-error, each arrangement suggesting new possibilities while also posing new questions. This aspect of the Kit-of-parts approach is one of its key advantages; it facilitates decision-making by introducing limitations, thereby eliminating what's not possible and thus suggesting a more narrow range of potential solutions that can be explored.

的初学者而言，这往往是最难跨越的一个挑战。仅依靠绘图来尝试三维层面的工作是非常局限的。在此，直观的模型操作是设计过程的关键，这是"从做中学"教学思想的本质体现。

"设计过程观"是另一教学法目标，在设计思维中培养一种方法化的过程是至关重要的。建筑设计，究其本质是一个过程——非线性的、结合个人的直觉与客观推演。在这一过程中，需要对于设计的可能进行反复的自我评价，通过自我设问讨论不同的思路。当然也不排除一些随机，甚至古怪的想法，但它指向的是一种"成败不定"的设计路径。在学习的过程中，重视记录不同的设计想法，并以此展开多方案间比较。在设计过程中，这是十分必要的，学生可以根据预设的一些目标来甄别出不同方案之间的优缺点，也可以根据这个目标对于决策进行相应调整。这是一个不断优化的过程，在设计的发展过程中维持一定的连续性与一致性，而非反复的否定。

在设计练习中，学生可选用一些柔软易处理的模型材料（如卡纸、泡沫板等）来制作研究模型，并用蓝丁胶、大头针对其暂时性的固定。这一做法方便学生对设计进行不断的调整，并通过拍照的方式记录不同阶段的成果。同时，允许学生在设计过程适时的回到前一步，再开始对于新方向的探索。从这个角度来看，制作研究模型可使得整个设计过程透明化，在指向多种设计可能的同时，持续的思考与改进设计。

装配部件教学法的第三点特点在于其关联抽象式形式构成与建筑设计的潜力。如前文所述，装配部件所使用的要素都是一些简单的抽象物件（杆件、板片、体块），它们所暗指的皆是建筑的组成部件（柱、楼板与墙体、剖碎等）。在包豪斯的设计教学中，抽象性同样是其诸多基础设计练习所共通的重要特征［图4.2-9］。但是，它们并不以建筑作为参照。在练习中使用铁丝、金属板、透明玻璃片等材料是为了创造出带有现代形式语言（如动态的不对称、透明等）的物件，而不是探索建筑空间或形式。

应该说"九宫格问题"是第一个同时探索"抽象"并指涉建筑的设计练习，以"九宫格问题"为起点，海杜克设计了一系列的住宅方案。在回顾了得州大学的"九宫格问题"习作后，海杜克在库伯联盟建筑学院的基础教学中重新展开了一套"九宫格问

The knowledge content of abstract formal composition includes many of the concepts introduced by Hoesli, Rowe, et. al. that address space perception and space design. Critical among these is the perception of space as form, having both shape and dimension. The interpenetration of spaces and the concept of phenomenal spatial transparency add a theoretical basis to an understanding of modern architectural composition. The distinction between figure and ground further clarifies our understanding of object and field, and informs the formal language of spatial composition. These concepts can be introduced and explored in a series of successive, issue-focused design exercises.

4) Pedagogical characteristics

There are several important pedagogical features imbedded in a design process that are enhanced by a kit-of-parts approach. First is the emphasis on *intuitive operation*. Direct manipulation of pre-configured model elements is encouraged in the kit-of-parts approach. By playing with the object elements, students benefit from both a tactile and visual relationship with form. Interaction with form in this way has the advantage of immediate three-dimensional visualization of potential configurations, overcoming one of the biggest challenges of spatial design for a beginning student. Using drawings alone to work in three-dimensions is an unnecessary limitation. Direct modelling is key to the design process. This is the essence of "learn from play".

Design as a process is a second pedagogical objective. The cultivation of a methodological approach in design thinking is critical. Architectural design is, after all, a process; non-linear and combining intuition with objective reasoning. Iterative reflection and revision of hypothetical "what if" design propositions are synonymous with the creative process. But random and idiosyncratic generation of alternative configurations results in a "hit or miss" approach to a design. Documentation or archiving of alternative proposals is necessary in order to compare results and identify advantages and disadvantages of each design proposal with respect to a prescribed set of objectives. Decision-making can thus be justified on an objective basis. This is a refining process in which continuity and consistency of a design development is preferable to constant negation.

In design exercises that use soft materials (cardboard, foam core, polystyrene foam, etc.), rubber cement and pins may be used to fix study models temporarily, providing flexibility to make changes. A photographic record of each stage is recommended. This technique allows the design process to return to a previous iteration in order to explore a different direction. The use

图 4.2-9
均衡研究，铁丝与铝片，
沃纳·齐默尔曼，
约1924年；
equilibrium study with
steel and aluminum,
Werner Zimmermann,
circa 1924;

of study models in this way supports a transparent and open-ended design process that can be continuously questioned and re-examined.

The third important feature of the kit-of-parts approach is the potential to relate *abstract formal composition and architectural design*. As previously described, the elements in each kit-of-parts are simple abstract objects (sticks, slabs and blocks) that allude to architectural components (columns, floors and walls, poché elements, etc.). In the Bauhaus method of design education, abstraction was an important feature in many of the basic design exercises [fig. 4.2-9]. However, it was not the intention to make reference to architecture. Exercises incorporating wire, metal plates, transparent fragments of glass and other materials, created objects that embodied the language of modern form (dynamic asymmetry, transparency, etc.) but were not necessarily explorations of architectural space or form.

Perhaps the first design exercise to explore abstraction with direct reference to architecture was the Nine-Square Problem. Following his series of theoretical house designs and revisiting earlier student exercises at UT that incorporated a nine-square grid, John Hejduk formulated the Nine-Square Problem as a beginning design exercise at The Cooper Union School of Architecture. In this problem, abstract formal elements (cube, cylinder, round column, slab wall, etc.) acquired architectural meaning in their deployment within the fixed nine-square grid-frame that served as site and context. Slabs in vertical orientation positioned themselves as freestanding partitions, directing and defining space. Cube and cylinder acted as freestanding objects in space, solid volumes easily imagined as service poché.

In the kit-of-parts approach represented by the program described in this book, architectural reference is introduced in two ways. First, operational constraints on the various components employed in the exercises respond to architectural issues, such as structure, space definition and human scale. For example, a portal opening in a vertical plane requires a lintel of a certain depth. Second, and more explicit, the lecture component of the course introduces the language and principles of form with reference to architectural precedent. Selected readings further reinforce key concepts with additional references and images.

题"——在作为不可变的九宫格网格框架（即场地）中，布置包括立方体、圆柱体、长圆柱、板片等在内的抽象的形式要素。在这个练习中，抽象的要素获得了建筑的意义：垂直摆放的板片成为了独立支撑的分隔墙，它引导并限定了空间；立方体与圆柱体成为空间中独立支撑的物体，其实体体量可以理解为剖碎。

本书所介绍的课程训练同样采用了装配部件设计教学法，并通过两种方式来引入关于建筑的参照。第一，练习中对于部件要素的操作限制对应于诸多具体建筑问题，包括结构、空间限定与人体尺度。比如说垂直板片上的洞口尺寸有考虑到其上过梁的高度。第二，采用一种更加明确具体的方式——通过课程讲座，结合建筑先例来讲解形式的语言与原则。同时为学生提供一些指定阅读材料，通过额外的参考资料以及图片来强化对于关键概念的认识。

5. 结语：建筑设计基础是否可教？

本文的讨论主要集中于一个重要且非常基本的问题：设计是否是一个可教的技能？我们的答案是肯定的。如何更好地教授设计对于今天的建筑教学仍是一个挑战。装配部件教学法在本质上是一个工具，它提供了一种鹰架式[16]的设计过程，它所代表的是一类教学思路，这其中包括（1）明确清晰的理论观点（即空间构成）；（2）由一系列周期短、问题明确的设计练习组成的一个循序渐进的学习平台；（3）一种设计目标明确的形式创造设计方法。

5. Conclusion: Are the fundamentals of design teachable?

Our discussion here centers on an important and very basic question: is design a teachable skill? We believe it is. How best to teach design remains a challenge in architectural education today. The kit-of-parts approach is essentially a device that provides a type of scaffolding structure[16] to the design process. It is part of a larger teaching agenda that includes (1) an

对于学生而言，装配部件教学法可以减少他们对于设计之神秘的担忧。它提供了一个设计的开启方式（"从做中学"的思想）以及一套基于任务书中所列明的设计目标的设计决策评价方法。事实上，该教学法中使用的部件要素的自身性质已经为学生预设了一套评价标准，并指引学生在设计的过程中进行决策。对于教师而言，这样一种标准化、系统化的设计方式消除了不同教师之间的风格差异，避免了对于同一份学生作业评价时可能出现的冲突与分歧。

毫无疑问，设计基础的教学方式多种多样。装配部件教学法的使用同样存在一定的问题，这个与抽象的程度，针对设计所预设的限制条件有关。但是，我们还是相信，装配部件教学法作为一种教学工具，它提供了一种理想的策略，使得建筑设计的初学者可以参与到对于空间以及空间与建筑设计之本质关系的讨论中。

explicit theoretical position (spatial composition), (2) a progressive learning platform that incorporates a sequence of short, issue-oriented design exercises, and (3) an approach to design that combines formal invention with explicit design objectives.

For students, the kit-of-parts approach can help reduce some of the mystery of design. It offers a way to get started (through the concept of "learn from play") and a means to assess design decisions based on the objectives outlined in the brief. The properties of the elements of the kit-of-parts provide a built-in set of criteria that guides decision-making throughout the process. For teachers, the standardization and systematic approach to design eliminates stylistic eccentricities among the instructors that otherwise present conflicting and divergent assessments of individual student work.

Certainly there are other viable approaches to the teaching of design fundamentals. Adopting a kit-of-parts approach has certain drawbacks associated with the degree of abstraction and the limitations placed on the creative design process. None the less, we have a strong conviction that as a teaching tool, the kit-of-parts approach offers an ideal strategy to engage beginning design students in a discussion about space and its essential relationship to architectural design.

注解 notes

1. "The IKEA Concept: Doing it a different way". See: https://www.ikea.com/ms/en_HK/this-is-ikea/the-ikea-concept/index.html.

2. Biber, George Eduard, and Jeffrey Stern. *Henry Pestalozzi, and His Plan of Education.* Bristol; Taipei: Thoemmes Press ; Unifacmanu, 1994.

3. "Froebel's Kindergarten Curriculum Method & Educational Philosophy". See: http://www.froebelgifts.com/method.htm

4. Rainer Wick, Gabriele D. Grawe, and Stephen Mason, eds. *Teaching at the Bauhaus.* Ostfildern-Ruit: Hatje Cantz, 2000: 102-104;

5. Liebschner, Joachim. *A Child's Work: freedom and Play in Fröbel's Educational Theory and Practice.* Cambridge: The Lutterworth Press, 1992.

6. Frank Lloyd Wright. *A Teastament.* New York: Brmhall House, 1957: 20.

7. Timothy Love. Kit-of-parts Conceptualism: Abstracting Architecture in the American Academy. *Harvard Design Magazine,* no.19 (2003): 42.

8. Ibid.

9. Durand, Jean-Nicolas-Louis. *Précis of the Lectures on Architecture: With Graphic Portion of the lectures on Architecture.* Getty Research Institute, 2000.

10. Alexander Caragonne. *The Texas Ranger: Notes from an Architectural Underground.* Cambridge, MA: The MIT Press, 1995: x.

11. Ibid., 190.

12. Ibid., 278-279, 302-305.

13. Jansen, Jürg. *Architektur lehren: Bernhard Hoesli an der Architekturabteilung der ETH Zürich = Teaching architecture : Bernhard Hoesli at the Department of Architecture at the ETH Zurich.* Zürich: GTA. 1989.

14. Kramel, Herbert. *Grundkurs 85: chronologische Darstellung der vier Arbeitsphasen des Grundkurses 1985 = Basic design program : chronological description of the four phases of the Basic Design Courses, 1985.* Zürich: GTA. 1986.

15. Friedman, Jonathan Block. *Creation in Space: A Course in the Fundamentals of Architecture,* Volume 1: Architectonics. Kendall/Hunt Publishing Company, 1999.

16. Beed, Penny L., E. Marie Hawkins, and Cathy M. Roller. "Moving Learners toward Independence: The Power of Scaffolded Instruction." *The Reading Teacher* 44, no. 9 (1991): 648-55. http://www.jstor.org/stable/20200767.

参考文献 references

1. Love, Timothy. Kit-of-parts Conceptualism: Abstracting Architecture in the American Academy. *Harvard Design Magazine,* no.19 (2003): 40-47.

2. Rubin, Jeanne S. "The Fröbel-Wright Kindergarten Connection: A New Perspective." *Journal of the Society of Architectural Historians* 48, no. 1 (1989): 24-37.

3. Caragonne, Alexander. *The Texas Ranger: Notes from an Architectural Underground.* Cambridge, MA: The MIT Press, 1995.

4. 布鲁斯·埃里克·朗曼, 徐亮, 顾大庆. 空间练习之装配部件教学方法. 《建筑师》172: 39-49.

插图出处 image sources

p218   Ulrich Franzen, Alberto Pérez Gomez, *Education of an Architect: A Point of View, the Cooper Union School of Art & Architecture.* Monacelli Press, 1999.

p219-1   Notes for *Basic Design 87-88* at ETH, provided by Gu Daqing.

p219-2   Friedman. *Creation in Space,* 8.

4.2-1   Daniel Dasey. "Thinking outside the box", 2016. See https://www.ikea.com/ms/en_KR/this-is-ikea/ikea-highlights/Flat-packs/index.html

4.2-2   Joseph Paxton, Charles Fox, and John McKean. *Crystal Palace: Architecture in Detail* (London: Phaidon Press, 1994), 13, 53.

4.2-3   Marilyn Neuhart and John Neuhart, eds., *Eames house* (Berlin: Ernst, 1994), 26.

4.2-4   Caragonne, The Texas Ranger: 194.

4.2-5   Jansen. *Bernhard Hoesli.* 50.

table 2   Friedman. *Creation in Space,* 33, 40, 54, 66 (up-left to up-right), 27, 41, 53, 63 (down-left to down-right)

4.2-7   Friedman. *Creation in Space,* 66.

4.2-8   Jürg. *Teaching architecture,* 50.

4.2-9   Wick, Gabriele, and Mason, eds. *Teaching at the Bauhaus,*158.

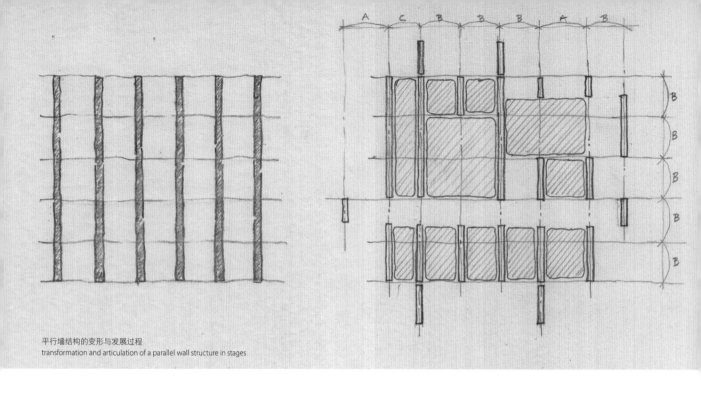

平行墙结构的变形与发展过程
transformation and articulation of a parallel wall structure in stages

徐亮 Xu Liang

4.3 研究 study
平行墙的形式操作 formal operation on parallel walls

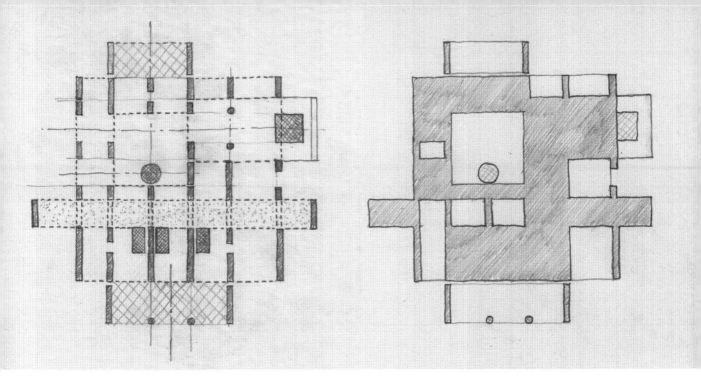

平行，强调沿单一方向的复制，是一种基本而有效的形式控制方法。通过将两组平行系统相互垂直叠加，可以衍生出一个网格系统。

在建筑中，平行系统的应用可初略地分为两类。其一是纯粹将平行作为一种形式策略，如路易斯·康设计的金博尔美术馆（1972）与伦佐·皮亚诺设计的贝耶勒基金会博物馆（1997）。而另一种方式是赋予平行线材料性，将其变为一组具有结构意义的线性墙体。换言之，建立平行墙体系。

本文的研究由两部分组成：其一，以图解的形式，对平行墙系统的形式操作过程进行抽象演绎，以此来展示空间限定要素的组织之于空间感知的影响；其二，我们选取了一系列建筑案例，从形式操作、材料、结构与建造的角度进行讨论，以检视平行墙体系的实际应用。

Parallel structure is a fundamental but efficient formal device by creating a repetitive pattern along a single direction. We can also reduce a grid into two perpendicularly overlapped groups of parallel lines.

In architecture, the application of a parallel system takes place in two ways. One regards parallel structure as a significant formal order, which can be observed in the Kimbell Art Museum (Louis Kahn, 1972) and the Foundation Beyeler (Renzo Piano, 1997). While the other approach further materialises the parallel line into an array of linear walls with certain structural contribution. In other words, to establish a parallel wall system.

This study starts from a diagrammatic speculation — the formal operation of an abstract parallel wall system, which aims to visualise the relationship between the space defining elements' configuration and one's spatial perception. Then we refer to a series of precedents that exemplify the adoption of such a system from the perspective of formal organisation, materials, structure and construction.

基本要素, wall elements

单一要素的重复, repetition of singular element type

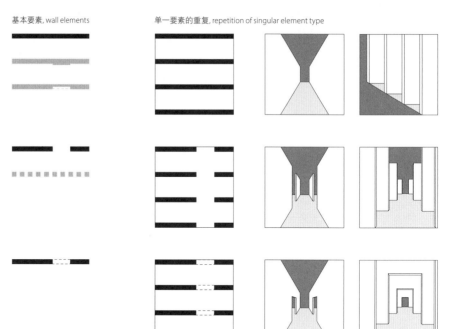

多种要素组合, combination of different element types

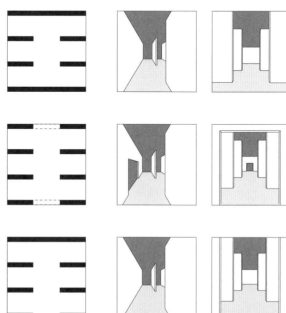

操作 – 重复，组合与调整
operation - repetition, combination and modification

　　基本墙体要素可分为三种: 连续墙，墙体片段（墙体间存在间隔）以及含洞口的墙体。在排布墙体时，可将单一要素重复布置，亦可将不同要素进行组合。当墙体之间的距离保持一致时，平行墙之间的空间是均质的。可以通过调节墙间距来获得层级性。移除部分墙体片段或开挖洞口可以连接这些彼此独立的空间。需要注意，当墙体片段之间的间隙过大时，对平行墙的感知会变弱，甚至消失。

　　There are three primary types of wall element: continuous wall, wall segments (wall with full-height breaks), and wall with openings. To establish a parallel wall system, we can repeat singular type of wall, or combine different wall types. When the distances between the parallel walls are equal, the whole wall system is rather neutral. Differentiating the in-between distances can establish a hierarchical system. We can subtract wall fragments or openings to connect these isolated spaces. However, the width of the wall breaks should be limited so as to retain the perception as a wall system.

调节尺寸, modify dimensions

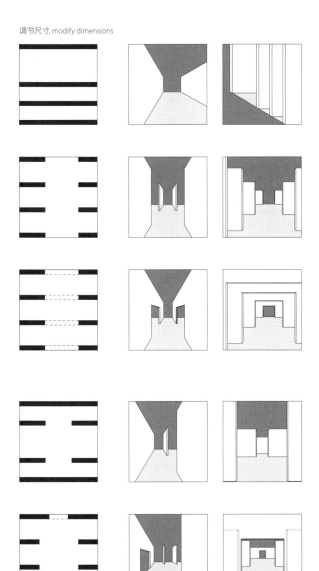

对平行墙的感知渐弱, the perception of parallel system is disminished

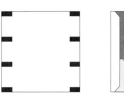

移除部分墙体, remove wall fragments

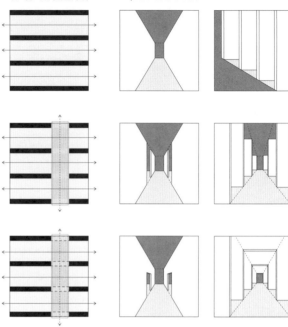

引入另一方向的空间带, introduce a space in other direction

空间关系随要素组合而复杂化, conbination of elements leads to spatial complexity

空间阅读 – 横纵空间限定
reading space - cross-grain spatial definition

空间操作对空间感知的影响可以从3个方面讨论：空间限定、空间的方向性与空间关系。当墙体相平行布置时，墙面之间限定出的线性空间强调的是单一的纵向性。通过挖去操作，可以创造多片墙体上相互对位的一系列开口，即引入正交方向上次一层级的空间带。在平面与透视图解中，我们可以通过连接这些开口的边缘与角落来可视化这些限定的空间体量。这些体量有着独特的尺寸、比例与朝向。

Space definition, directionality and spatial relationship are the key issues in the discussion of spatial operation and perception. The linear space defined by two parallel walls emphasises the longitudinal direction. By creating aligned openings in the walls through the process of subtraction, a secondary cross-grain spatial matrix can be generated in the transverse direction. In the plan and perspective diagrams, we can connect the edges and corners of the openings to visualise the defined volumetric space that has specific dimensions, proportions and orientation.

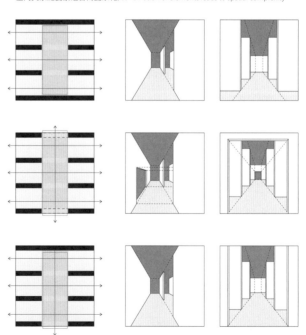

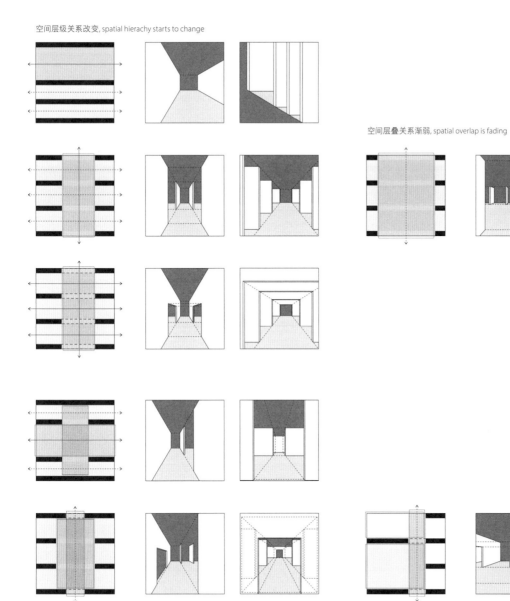

空间层级关系改变, spatial hierachy starts to change

空间层叠关系渐弱, spatial overlap is fading

239

屋顶 – 水平向的空间限定
roof - introduce horizontal definition

　　一方面，屋顶作为一种水平向的空间限定要素，其遮蔽属性可以有效地区分室内与室外。另一方面，在形式操作中，水平空间限定要素的布置可加强或改变原有的空间层级关系。右侧图解讨论了双边支承的矩形屋顶对于平行墙体系空间关系的影响。由于支承的构造要求，屋顶多覆盖于平行墙之间的区域，这一部分的空间随围合程度的加强而表现为实体，实体与虚空、室内与室外之间的对比使得空间的阅读与体验更加丰富。

　　Roof elements may be used to differentiate interior and exterior space by providing horizontal shelter. In formal operation, the arrangement of horizontal elements can further strengthen or alter the spatial hierarchy. The diagrams on the right illustrate how the roof with supports along the two opposite edges reinforces the reading of spatial relationship. The covered (solid) area, compared to the uncovered (void) ones, is further enclosed. The contrast between solid and void, interior and exterior allows a richer spatial reading and experience.

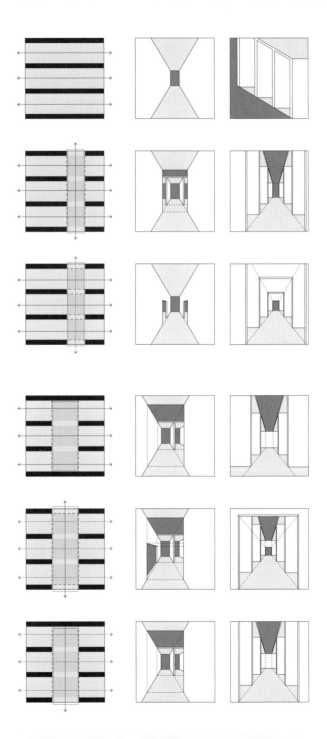

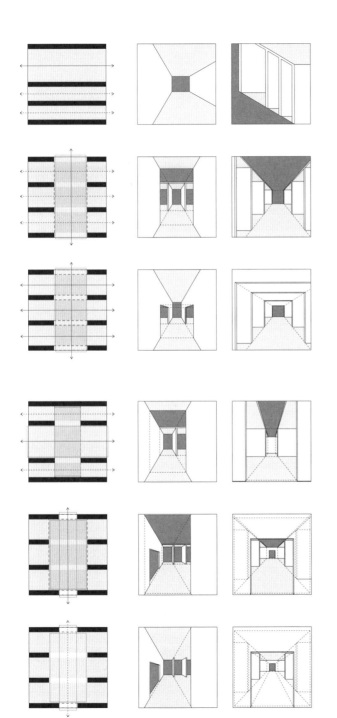

水平限定策略的影响, different strategies on horizontal definition

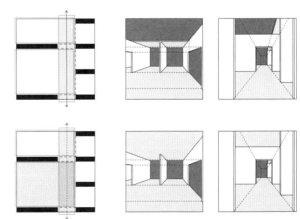

241

平行墙体系作为一种形式基准及结构类型，被广泛应用于当代建筑实践中。作为香港的第一批公营房屋计划，兴建于1954年的石硖尾徙置大厦采用的便是钢筋混凝土的平行墙结构，其突出优点在于经济性与施工便利性。

在现代建筑中，针对平行墙体系的探索多种多样。下文选取了一系列典型案例，通过对其进行分析以讨论平行墙体系中空间形式组织与材料、建造之间的关系。

荷兰建筑师阿尔多·凡·艾克（1918-1999）设计的桑斯比克雕塑展亭（1965-1966）是平行墙案例中最为人所熟知的一个经典作品。建筑主体是6片相互平行、等距排布的承重砖墙 [图4.3-1]，墙体上部支承水平钢梁及屋顶。展亭所在的场地三面环树，凡·艾克有意识地将平行墙的一侧实体墙面放置在较为开阔的第四面，面向公园中游客到来的方向。这样"结构就不会过多地暴露其中的内容，游客需要足够靠近，从两侧接近它"[1]。平行墙的两侧边（由墙体尽端边缘组成的面）与实墙面之间的虚实差别，是平行墙体系的一个突出特点。

通常，在平行墙体系的基本模型中，场地被平行墙体分隔为一组相互独立的纵向空间，建立墙体两侧空间之间的联系是获得空间丰富性的手段之一。在雕塑展亭中我们可以观察到3种基本操作：开挖门洞、挖去通高墙体片段与弯曲墙面。墙体之间布置有雕塑展示以及供游客休息的坐台。从之后展出的方案草图中，我们可以推测建筑师设计过程中的一些思考，其中平行墙的数量、半圆形墙与门洞的组合都有不断的

As a formal datum and structural type, the parallel wall system is widely used in the practice of contemporary architecture. The first public housing program in Hong Kong, Shek Kip Mei Resettlement Estate (1954) adopted the reinforced concrete parallel wall structure for its efficiency and constructional advantage.

In this section, a group of precedents are selected and analysed to address three issues — spatial organisation, materials and tectonics — with reference to the exploration of the parallel wall system in modern architecture.

Designed by the Dutch architect Aldo van Eyck (1918-99), the Sculpture Pavilion at Sonsbeek (1965-66) is a representative case: six self-supporting masonry walls, the primary elements of the pavilion, are arranged parallelly and equidistant. [fig. 4.3-1] Horizontal steel beams span between the top of the walls and supports the roof panels. The pavilion was located within a park with its three sides surrounded by trees and bushes. Aldo van Eyck placed one of the solid sides facing the fourth side, where visitors would come from. In this way, "the structure should not reveal what happens inside until one gets quite close, approaching it from the (wall) ends" [1]. In this regard, the architect intentionally used the contrast between the two sides of a group of parallel walls, one consisting of wall ends and the other a solid surface, the primary formal characteristic of the parallel wall system.

In general, parallel walls divide the field into a series of longitudinal zones. Establishing connection and spatial overlap between these zones contributes to a spatial complexity. In this pavilion, there are three primary types of wall operations: create portals, subtract sections of the wall and bend the wall. Apart from the walls, stands for sculpture display and benches for resting were inserted. From the published sketches, we can trace the architect's design thinking and process, during which both the

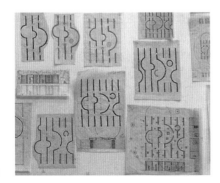  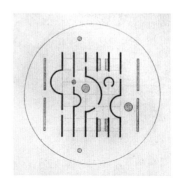

图 4.3-1

方案草图及平面，
桑斯比克雕塑展亭；
sketches and final plan,
Sculpture Pavilion;

调整。同样值得注意的是草图中绘制的辅助线，其中一类表达的是联系不同开口的视线或流线关系，它是确定平行墙上不同开口位置及相互关系的凭据之一；另一类辅助线是垂直于平行墙的一组平行线，这些线条之间的距离与平行墙间距相同，与平行墙形成了正交网格系统。在最终方案的平面中，网格线确定了所有的形式操作，包括定位门洞、物件摆放、半圆形墙的圆心及半径等。

同样是室外展览建筑，在瑞士建筑师彼得·卒姆托（1943-）设计的瑞士展览馆——声音盒（2000）中亦使用了平行墙体系。在这个大体量的展览馆中，建筑师将12组高9m的平行墙［图4.3-2］布置在51.2m x 58.3m的沥青垫上。这些平行墙可根据其形式操作分为两类：一类是沿单一方向重复的墙体系统，强调的是平行墙之间的通道性空间，共4组分列于场地四角；另一类则是将平行墙中央位置的部分墙体片段移除，以获得一个位于平行墙内部的小空间。

在这个项目中，另一个值得注意的问题是平行墙组之间的组合方式。平面上有一系列夹于实体之间的矩形开放空间，4组大小比例不同的平行墙系统围绕着这些矩形空间呈风车状布置。这些空间按照尺度及形态特征可以分为两种，一种是小尺度的，矩形各边可以向各平行墙组的内部渗透，称为"十字院"；另一种的尺度较大，矩形空间的各边由平行墙的实体面限定，中间置有3层高、呈放射形的服务单元，它们被称为"侧院"。

该方案的另一突出特点在于平行墙墙体的结构构造。平行墙墙体是由标准截面为100mm x 200mm，长

number of parallel walls and the combination of semi-circular walls and openings were adjusted. It is important to note the referential lines in these sketches. They were used to identify the direction of views or circulation, through which to determine the location of openings and their relationship. The other reference lines were parallel and perpendicular to the linear walls. In this way, a grid system with equal distances along the two directions was established. As presented in the final plan, the grid defined the arrangement of doorways, sculpture objects and the semi-circular wall fragments.

Swiss Sound Box, Swiss Pavilion (2000), designed by Peter Zumthor (1943-), is the other example of a parallel wall system for exterior exhibition use. Twelve groups of 9m tall parallel walls [fig. 4.3-2] were arranged on 51.2 x 58.3m asphalt cushions. In terms of formal operation, the wall groups can be divided into two categories. In four groups of walls, the linear wall elements were multiplied in a single direction and the tunnel space between each two walls was emphasised. These occupy the four corners of the site. In the other wall groups, wall fragments at the centre were subtracted to achieve a small area for rest.

In this project, the other issue to be noticed is the combination of wall groups. Four wall groups were arranged around an open courtyard in a pin-wheel pattern, a rectangular void subtracted from the dense covered fabric of the solid. The smaller squares were called "cross courts" with their four sides defined by "end walls". The larger squares were named "flanked courts", which are defined by the solid wall surfaces.

The other distinct feature of this project is the wall construction. Timber sticks of 100 x 200mm rectangular section and various lengths were stacked together to form the walls. Two types of wood, Larch and Douglas Fir, were used and arranged according to the wall directions, north-south and east-west. In contrast to conventional practice, the timber elements

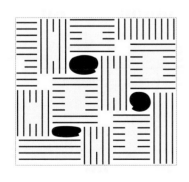
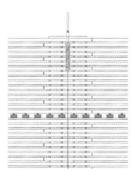
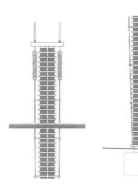

图 4.3-2
平面及剖面，
瑞士展览馆—声音盒；
plan and sections,
Swiss Pavilion;

度不定的木条层叠而成。墙体木材分两种，正南北向的墙体为落叶松，东西向的墙体为花旗松。展览期间，木条会发生自然的风干作用，由此带来的收缩会通过钢索连接的弹簧拉实，以保证整体结构稳固。等距排布的平行墙体之间由过梁连接，其上铺设预制金属槽板形成屋顶。

勒·柯布西耶（1887-1965）设计的萨拉巴依别墅（1951-1955）[图4.3-3]是平行墙体系探索的另一代表作。项目位于印度艾哈迈达德，建筑师在平行墙的布局与朝向上有考虑热带气候下对于通风的需求。平行墙的主体结构在高度上分为两段，下方是平行排布的砖砌墙体片段，上部支承混凝土连续过梁，过梁上通过起筒形拱顶形成水平向联系。砖墙的表面处理分两种：表面上刷白色或其他颜色抹灰，或是直接暴露砖砌构造。室内外之间的气候边界多采用木质隔断，在材料上明显区别于平行墙的结构主体。

支承过梁的砖墙在满足力学要求的同时，可相对自由变化其长度及对位关系，以创造出连续、流动的空间状态。其中，平行墙体间等宽的纵向空间带与穿越各墙体的横向空间带相互层叠，表现出一种现象的空间透明性，这是柯布西耶作品的突出特点之一。在这个设计中，家居使用功能被布置在这一些宽度不同的横向空间带中，连续的拱形空间带则向短边方向延伸并强化空气流动。

在鲁道夫·辛德勒（1887-1953）设计的洛弗尔海边住宅（1922-1926）[图4.3-4]中，其平行墙结构的形式与功能明显区别于前面介绍的几个案例。首先，他利用混凝土剪力墙的材料特点，创造出不规则形的墙

were allowed to dry naturally after assembly. At the same time, steel cables with springs compressed the wall structure to ensure its steadiness. The equidistant walls were bound by timber beams, on which prefabricated metal channels were placed to form the roof.

Villa de Madame Manorama Sarabhai (1951) by Le Corbusier (1887-1965) [fig. 4.3-3] is different from the aforementioned two cases. The building is located in Ahmedabad, India, where the tropical climate requires consideration of natural ventilation that is achieved in the parallel walls' layout and orientation. The main body of the wall structure can be divided into two parts. Parallel brick wall fragments support continuous concrete beams on which barrel vaults span across the walls providing horizontal structural support for either a roof or floor above. The treatment of wall surfaces varies — painted with white or coloured plaster, or just exposed rough coursed brick masonry. The climate border consists of pivot doors with veneer wood finish materially differentiated from the primary structure.

The brick walls in various lengths are necessary for load transfer. Meanwhile, these wall fragments are configured so as to establish a gridded open plan within which transverse spatial zones of varying width are defined. The overlapping of the vaulted longitudinal spaces defined by the parallel walls and the transverse spaces defined by the gaps in the walls suggests a type of phenomenal spatial transparency that is characteristic of the work of Le Corbusier. The program of the house is accommodated within the dimension of the transverse spaces while the continuous, uninterrupted vaulted spaces extend through the short dimension of the house enhancing air movement.

With regard to the form and function of the parallel walls, the Lovell Beach House (1922-26) [fig. 4.3-4] by Rudolf Schindler (1887-1953) is unique. Benefitting from the advantages of reinforced concrete, Schindler created

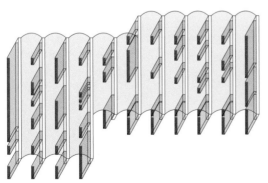

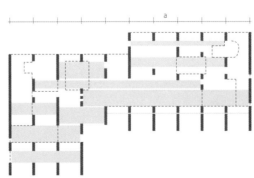

图 4.3-3

分析图解，
萨拉巴依别墅；
analytical diagrams,
Villa de Madame
Manorama Sarabhai;

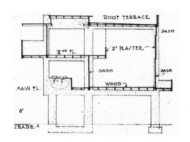

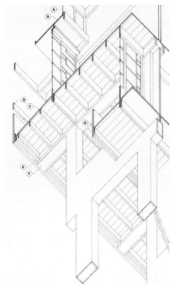

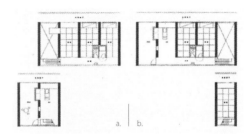

a. | b.

five equally spaced parallel irregular wall frames cast-in-situ. Timber beams inserted between the frames carry the timber floor and roof panels. In this way the materials further differentiate the structural systems of span and vertical support.

In this project, the architect carefully configured the wall frames in order to articulate the spatial relationship in the section. The two openings on the left and right corners respond to the two primary interior spaces — the double-floor living room on the first floor and the bedroom on the second floor. The floor surface on the second floor is extended to create an interior corridor and an exterior balcony. The last opening on the first floor is connected to the ground floor entrance with staircases.

Adopting reinforced concrete parallel walls as the primary structural elements is quite common in contemporary collective housing projects. Gifu Kitagata Apartment Building (1994-98) designed by Kazuyo Sejima (1956) is an example. From the typical floor plans [fig. 4.3-5], wall fragments are subtracted to accommodate in-house vertical circulation, horizontal corridors, and service elements.

The Multimedia Workshop in Oogaki (1997), designed by SANAA (Sejima & Ryue Nishizawa), demonstrates a mixed-use of materials for the parallel wall system. In the plan [fig. 4.3-6], a group of eight parallel walls is centred in the field with corridors surrounding its four sides. Reinforced concrete and timber are the two materials used in the wall construction. Four reinforced concrete parallel walls are the primary load-bearing elements, while the other four timber walls are mainly used for spatial division. The architects vary the distances between the eight walls responding to different programs, e.g. studio, ateliers, passage, living room, storage, etc. In this way, a group of rooms with spatial hierarchy was established mainly based on programmatic requirement. In the meantime, since these programs are independent from each other, the design did not emphasise the connection between the two sides of a wall. People can only cross the zones through the openable doorways.

体，其上开挖多个墙洞而表现为"墙框"。然后将这5片现浇的平行墙框等距离排布，在墙之间夹木梁，上铺木板以形成楼面及屋面，从材料上区分出水平联系与垂直传力的构件。

在这个项目里，建筑师通过设计这些墙框来组织剖面上的空间关系。建筑的室内部分主要包括两层通高的起居室以及位于三层的卧室，这两部分直接对应于平行墙框上的两个洞口。三楼的楼板向内延伸形成伸入起居室的内走廊，向外延伸形成阳台。二楼的另一墙洞则通过楼梯与地面层的主入口联系。

将平行剪力墙作为主要的结构构件，是当代集合住宅设计的一个普遍现象。由日本建筑师妹岛和世（1956-）设计的岐阜公寓（1994-98）便是其中一例，项目中建筑师利用平行剪力墙的特点来组织户型

设计 [图4.3-5]，包括跃层交通，水平走道以及服务功能。

在SANNA（妹岛和世与西泽立卫）设计的大垣多媒体工作坊（1997）中，混合使用了不同材料来建造结构性与非结构性的平行墙。如平面图 [图4.3-6] 所示，方形平面的中央排布有8片平行墙，其周围是一圈回廊。在这8片平行墙中，有4片起主要结构作用的钢筋混凝土结构平行墙；其余4片墙体为木龙骨结构，主要用以划分空间。8片平行墙之间的距离不一，由此产生的层级关系对应于不同的使用功能，包括通道、工作室、客房、仓库等服务空间。由于各功能之间的独立性，平行墙两侧仅通过可开启的门洞联系，不强调空间之间的流动性。

从上文的分析中，我们可以看到，平行墙体系在空间形式组织中所表现出的灵活性。单一方向重复布置的垂直板片将空间分割为一组与墙体平行的等距纵向空间。改变墙体之间的距离是获得空间层级性的一种方式。另外，通过建立墙体两侧空间之间的联系——横向的空间带，可以创造出更加丰富的视线、流线及空间体验。

对于平行墙体系的建造，所使用的材料、结构构造与其他建筑体系并无多大的差别。需要重视的主要包括两方面：墙体上门洞开口及墙体间水平联系要素的构造方式。雕塑展亭中砌体门洞上方布置有过梁，萨拉巴依别墅中使用的是连续的钢筋混凝土通长横梁。在钢筋混凝土墙体中，因为无需使用过梁，墙上洞口的布置及尺寸则相对自由。平行墙体系中的墙体需要支撑来自水平跨度构件（上部楼板或屋顶）的荷载。如何连接这两种不同类型的要素是至关重要的。水平要素可直接支承在墙体顶部，这是一种相对普遍的搭接方式，或是如洛弗尔海边住宅一样将水平构件置于墙体之间。在钢筋混凝土结构中，也会将水平板片与垂直墙体现浇在一起。因为钢筋混凝土结构具有整体性的特征，这使得它适用于多层的平行墙体系，这也带来了垂直方向上丰富的空间关系。同时，刚性的水平联系可以加强系统的水平稳定性。

本文对于平行墙体系及其设计的研究突出强调其对于建造、结构与空间逻辑的整合，这是平行墙体系的重要特征。该系统不仅是一种强有力的形式机制，同时也控制建构的表现力及特征。

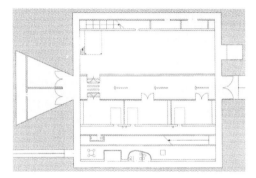

图 4.3-6
平面及墙体材料区分，
大垣多媒体工作坊；
wall material
differentiation in plan,
Multimedia Workshop in
Oogaki;

From the above analyses, we can see the flexibility of the parallel wall system with regard to spatial organisation. The walls repeating along one direction divide the field into a series of equal width longitudinal spatial zones with the space defined by the two solid wall surfaces. Varying the spacing of the walls is a convenient way to achieve spatial hierarchy. Furthermore, by creating transverse spatial zones, we can generate a rich spatial experience as well as obtain cross-sectional views and circulations.

In terms of material, structure and construction, there is nothing unconventional about the parallel wall system. Two things to be noticed are the construction of wall openings and the horizontal connection among the wall pieces. Lintels are used above the openings in the case of the Sculpture Pavilion, while continuous deep concrete beams are adopted in Villa Sarabhai. Walls of reinforced concrete wall do not require lintels, therefore the placement and width of openings is more flexible. In the parallel wall system, the wall elements support the loads of the horizontal spanning structure, the upper floor or roof planes. Connecting these two types of elements is a key question. Placing the horizontal elements on the top of the walls is one approach. Alternatively, the horizontal elements can be inserted in-between the walls, as seen in the Lovell Beach House. In reinforced concrete structure, horizontal planes may be cast together with the vertical wall slabs. The monolithic character of reinforced concrete lends itself to a multi-story parallel wall structure and facilitates the creation of a rich spatial relationship in a section. A strong and rigid horizontal connection contributes to the lateral stability of the system.

This study of parallel wall design emphasizes an important and distinctive feature of the system: that of the integration of constructional, structural and spatial logic. The system is not only a powerful formal device but also one possessing the potential of tectonic expression and character.

注解 notes

1. Vincent Ligtelijn, *Aldo van Eyck Works*, Birkhauser Publisher, 1999: 134.

参考文献 references

1. Zumthor, Peter, and Plinio Bachmann. *Swiss Sound Box: A Handbook for the Pavilion of the Swiss Confederation at Expo 2000 in Hanover*. Basel [Switzerland] ; Boston: Birkäuser, 2000.
2. Gargiani, Roberto, and Anna Rosellini. *Le Corbusier: Béton Brut and Ineffable Space (1940 - 1965): Surface Materials and Psychophysiology of Vision*. EPFL Press, 2011.
3. Sejima, Kazuyo, and Ryue Nishizawa. "Multimedia Workshop Oogaki, Gifu," *Lotus International*, no. 96 (1998): 24-31.

插图出处 image sources

4.3-1   Vincent Ligtelijn. *Aldo van Eyck Works*. Birkhauser Publisher, 1999: 141, 134.
4.3-2   Thomas Durisch, *Peter Zumthor 1985-2013: Buildings and Projects*, Volume 2 (1990-1997), Zurich: Scheidegger & Spiess. 2014: 120.
4.3-4   Judith Sheine and R. M. Schindler, R. M. Schindler, Coleccíon Obras y proyectos - Works and projects series (Barcelona: Gili, 1998): 67;
Edward R. Ford, *The Details of Modern Architecture*, Volume 1, Cambridge: The MIT Press. 1990: 302.
4.3-5   *El Croquis 99: Kazuyo Sejima + Ryue Nishizawa, 1995-2000* (1998): 42;
4.3-6   Kazuyo Sejima and Ryue Nishizawa. "Multimedia Workshop Oogaki, Gifu," *Lotus International*, no. 96 (1998): 31.

致谢 acknowledgements

本项研究部分得到香港特别行政区研究资助局（项目编号：CUHK443812，2013-14）资助。

在过去的几年间，我们在香港中文大学建筑学院开设了一门一年级基础设计课程"建筑学导论"。本书的写作是对该课程教学的记录与发展。在教学过程中，我们有幸和诸多优秀的导师合作：Samuel Diedering、耿炎、Pedram Ghelichi、叶懿乐、高家扬、廖晋廷、毛家谦、姚凯琳、孙宇璇、冯诗蔚、吴佳维、翟玉琨，以及学院工作坊的戴温明。另外，钟文慧、林颢谕、李俊铺、马廷光、黄乐希、吴宇商作为助教为我们提供了很多帮助。作为一个教学团队，我们共同为建筑系的新生以及大学范围内其他对建筑学感兴趣的学生提供关于设计入门的指导，在此感谢他们对于课程教学的贡献。我们同样感谢所有参与课程学习的学生，他们对于设计的好奇与热情以及出色的创意是我们持续教学研究的动力。

另外，我们要特别感谢本书的顾问顾大庆教授。顾教授对于基础设计教学法的研究与探索对于中国建筑教育产生了重要的影响。在本书的写作中，顾教授提出了诸多宝贵建议，感谢其长久以来的信任与支持。在此，我们还要感谢建筑学院的前院长陈丙骅教授（2014-2019）对于开设这门一年级入门设计课程的支持。

最后，我们希望感谢得州骑警这一批先驱教育家的遗赠以及一代又一代的后继者对于建筑教育中形式设计方法的持续探索。

The work described in this book was partially sponsored by the Research Grants Council of the Hong Kong Special Administrative Region (Project no. CUHK443812, 2013-14).

This book evolved from our teaching the first-year design course, *Introduction to Architecture*, at the School of Architecture at the Chinese University of Hong Kong. Over many iterations of the course, we have been fortunate to work with a number of talented and capable instructors including Samuel Diedering, Geng Yan, Pedram Ghelichi, Ip Yi Lok Ivy, Rina Ko, Liu Chun Ting Larry, Mo Kar Him, Yiu Hoi Lam Melody, Sun Yuxuan, Sze Wai Fung Veera, Wu Jiawei, Zhai Yukun and our school technician Dai Wan Ming Leo. In addition, Chung Man Wei Jenny, Lam Ho Yu Jacky, Li Jun Paul, Ma Ting Kwong Marcus, Wong Lok Hei Jessie, Wu Yu-shang Sunny have contributed as teaching assistants. Together as a team we have offered beginning design instruction to entering architecture majors and interested non-majors in various disciplines throughout the university. We thank all of the students who have participated in the course for their curiosity, enthusiasm and often outstanding creative work that has contributed to this pedagogical study.

We owe special thanks to Prof. Gu Daqing, our advisor, for his inspiration and continuous support. His long term leadership in the pedagogy of foundation design has been a significant influence on beginning design education throughout China. Also we thank Prof. Nelson Chen, Past Director of the School of Architecture (2014-2019) for his commitment and initial support of the introductory first-year design course.

Last, we must acknowledge the legacy of the early pioneers, the Texas Rangers, as well as the following generation of design educators who extended the vision of a formal design methodology in architectural education.

导师与学生合影，2015-2016学年
instructors and students, AY 2015-2016

模型制作工作坊
workshop on model making

作者 authors

布鲁斯·朗曼 Bruce Lonnman

朗曼于雪城大学及康奈尔大学完成其建筑学专业教育，并分别获得建筑学学士及建筑学（城市设计）硕士，他同时拥有康奈尔大学的土木工程学士及硕士学位。作为1983年第一届SOM旅行奖学金的获得者，他游历了位于法国、瑞士以及意大利的欧洲建筑。在经历了短暂的实践后，他开始了其建筑教育职业生涯，先后于肯塔基大学、俄亥俄州立大学、雪城大学以及佐治亚理工学院任教。2002年至2007年间，他于沙迦美国大学任教并担任系主任一职。

朗曼现为香港中文大学建筑学院兼任副教授，教授建筑设计及建筑结构相关课程。作为三年级"第四建筑设计工作室"的课程负责人，他在设计课程中引入建筑技术及响应气候设计作为训练主题。另外，他创立了以问题导向、抽象设计练习为主的基础设计课程——"建筑学导论"。朗曼的学术研究主要包括结构模型、建筑教育中的结构设计以及基础设计教学。

Bruce Lonnman completed his architectural education at Syracuse University (BArch) and Cornell University (MArch), where he also received degrees in Civil Engineering (BS and MCE). As a recipient of a Skidmore Owings and Merrill (SOM) Travel Fellowship in 1983, he visited and studied European architecture primarily in France, Switzerland and Italy. After a short period of time in practice, he began a career in architectural education, first at the University of Kentucky, followed by Ohio State University, Syracuse University, and Georgia Institute of Technology. From 2002 to 2007, he taught at the American University of Sharjah and served as Department Chair.

Currently an adjunct associate professor at The Chinese University of Hong Kong (CUHK), Lonnman teaches architectural design and building structures. As the Coordinator of the U4 third year Design Studio, he introduced building technology and climate responsive design to the Studio as major themes. In addition, he has created the foundation course, *Introduction to Architecture*, which offers studio-based learning through abstract, issue-driven design exercises. Lonnman's previous publications include articles on structural models, structural design in architectural education, and beginning studio design teaching.

徐亮 Xu Liang

徐亮现为香港中文大学建筑学院讲师，他于该校毕业并获得建筑学哲学博士学位。在此之前，他曾在内地接受土木工程与建筑学的专业教育，并参与实际建筑项目。现阶段他的教学工作主要包括一年级的基础设计（"建筑学导论"）、二年级的设计工作室（"建筑设计导论"一、二）及其他专业基础课程（"绘图与视觉设计"、"建筑学基本问题"等）。

自2013年起，徐亮参与了关于建筑教育与设计教学法的诸多研究课题，其中包括顾大庆教授主导的基金项目"设计教学法理论研究——练习、设计和教案"。他的博士研究讨论了基于叙事的设计教学法在中国建筑教育中的移植及其特征、影响。除了设计教育，他的研究兴趣还包括叙事与建构，建筑展览与策展，建筑摄影。

Xu Liang is a lecturer in the School of Architecture at The Chinese University of Hong Kong, where he graduated with a Doctor of Philosophy (PhD) in Architecture degree. Prior to this, he received his professional education in Civil Engineering and Architecture and practiced as an assistant architect in Mainland China. His current teaching includs the first-year basic design (*Introduction to Architecture*), second-year design studios (*Introduction to Architectural Design*, U1&U2) and other foundation courses (*Graphics and Visual Studies*, *Architecture Fundamentals*, etc.).

Since 2013, Xu has been involved in research projects focusing on architectural education and design pedagogies, e.g., "A Theoretical Study of Design Pedagogy: Exercise, Project and Program", led by Prof. Gu Daqing. His PhD research investigated the transplantation of a narrative-based design pedagogy in China and its impact on China's contemporary architectural education. Apart from design education, his current research explores narrative and tectonics, architecture exhibition and curation, architecture photography.

图书在版编目（ＣＩＰ）数据

抽象构成与空间形式 = Abstract Composition and Spatial
Form : 汉英对照 / (美) 布鲁斯·朗曼, 徐亮
著. -- 北京 : 中国建筑工业出版社, 2020.6（2023.8 重印）
　ISBN 978-7-112-25070-7

　Ⅰ. ①抽… Ⅱ. ①布… ②徐… Ⅲ. ①建筑设计－艺术构
成－汉、英 Ⅳ. ①TU2

　中国版本图书馆CIP数据核字(2020)第073992号

策划：顾大庆
责任编辑：滕云飞　徐　纺
责任校对：赵　菲

抽象构成与空间形式
Abstract Composition and Spatial Form
[美] 布鲁斯·朗曼　徐亮 著

*
中国建筑工业出版社 出版、发行〔北京海淀三里河路9号〕
各地新华书店、建筑书店经销
北京利丰雅高长城印刷有限公司印刷
*
开本：889毫米 X 1194毫米 1/20 印张：13²/₅ 字数：480千字
2020年11月第一版 2023年8月第四次印刷
定价：48.00元
ISBN 978-7-112-25070-7
　　　　　（35788）

Vernacular Contained
香港集装箱建筑
Vito Bertin · Gu Daqing · Jaffa Pak Lung Chong China Architecture & Building Press

Introduction to Architectural Design
建筑设计入门
Gu Daqing · Vito Bertin | China Architecture & Building Press

Space, Tectonics and Design
空间、建构与设计
Gu Daqing · Vito Bertin 著 | China Architecture & Building Press

leverworks: one principle, many forms
杠作：一个原理、多种形式
Vito Bertin | China Architecture & Building Press

Abstract Composition and Spatial Form
抽象构成与空间形式
Bruce Lonnman · Xu Liang | China Architecture & Building Press

《香港集装箱建筑》是我们出的第一本书，方开本，单一颜色纯文字的封面。后来出版的几本书都延续了相同的格式，自然就形成了一个系列。这个已经出版以及计划出版的书目清单突显了这个系列丛书的主题，即都与建筑设计教学密切相关，有些直接来自于所教授的课程，有些涉及建筑设计理论和方法的专门课题。所列出的计划出版书目绝大多数都在制作之中，将在几年内陆续完成，书名为暂定，以最后出版为准，个别课题或许会调整，还有新的计划加入。

已出版：

香港集装箱建筑 柏庭卫 顾大庆 胡佩玲 2004
建筑设计入门 柏庭卫 顾大庆 2009
空间、建构与设计 柏庭卫 顾大庆 2011
杠作：一个原理、多种形式 柏庭卫 2012
基础设计·设计基础 赫伯特·克莱默 顾大庆 吴佳维 2020
抽象构成与空间形式 布鲁斯·朗曼 徐亮 2020

待出版：

建筑设计入门之基本问题 顾大庆
建筑设计入门之基本方法 顾大庆
观察：日常建筑学 顾大庆
体验素描与设计素描 顾大庆
设计与视知觉（新版）顾大庆
建筑分析 柏庭卫
剖碎与透明 顾大庆
勒杜：九宫格立方体 顾大庆 陈乐
空间：从绘画到建筑 顾大庆
空间：从概念到建筑 顾大庆
模型之空间操作 顾大庆 柏庭卫 朱竞翔
教学法：练习、设计与教案 顾大庆
中国建筑教育的转折点 顾大庆
香港现代建筑 顾大庆 柏庭卫 韩曼

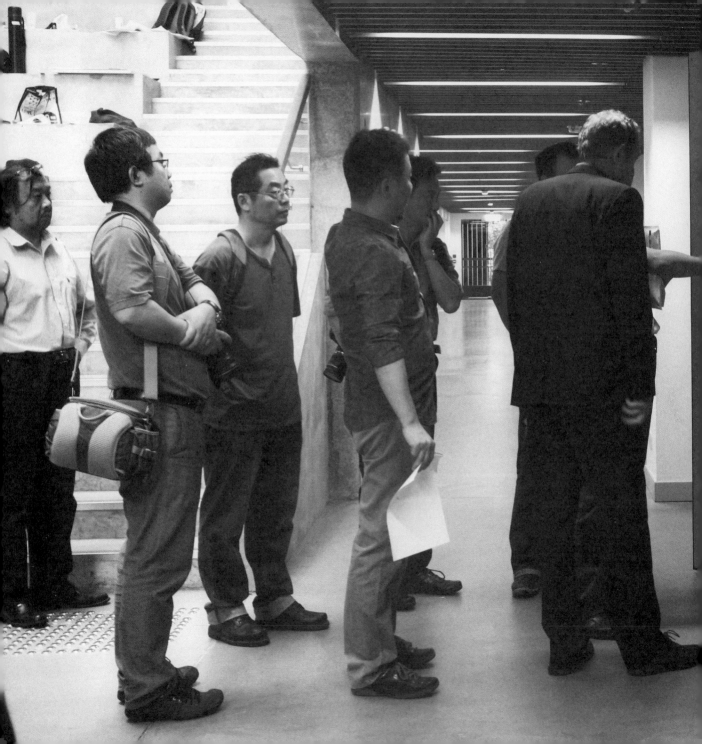